Textbooks in Telecommunication Engineering

Series Editor

Tarek S. El-Bawab

Professor and Dean

School of Engineering

American University of Nigeria

Yola, Nigeria

Dr. Tarek S. El-Bawab, who spearheaded the movement to gain accreditation for the telecommunications major is the series editor for Textbooks in Telecommunications. Please contact him at telbawab@ieee.org if you have interest in contributing to this series.

The Textbooks in Telecommunications Series:

Telecommunications have evolved to embrace almost all aspects of our everyday life, including education, research, health care, business, banking, entertainment, space, remote sensing, meteorology, defense, homeland security, and social media, among others. With such progress in Telecom, it became evident that specialized telecommunication engineering education programs are necessary to accelerate the pace of advancement in this field. These programs will focus on network science and engineering; have curricula, labs, and textbooks of their own; and should prepare future engineers and researchers for several emerging challenges.

The IEEE Communications Society's Telecommunication Engineering Education (TEE) movement, led by Tarek S. El-Bawab, resulted in recognition of this field by the Accreditation Board for Engineering and Technology (ABET), November 1, 2014. The Springer's Series Textbooks in Telecommunication Engineering capitalizes on this milestone, and aims at designing, developing, and promoting high-quality textbooks to fulfill the teaching and research needs of this discipline, and those of related university curricula. The goal is to do so at both the undergraduate and graduate levels, and globally. The new series will supplement today's literature with modern and innovative telecommunication engineering textbooks and will make inroads in areas of network science and engineering where textbooks have been largely missing. The series aims at producing high-quality volumes featuring interactive content; innovative presentation media; classroom materials for students and professors; and dedicated websites.

Book proposals are solicited in all topics of telecommunication engineering including, but not limited to: network architecture and protocols; traffic engineering; telecommunication signaling and control; network availability, reliability, protection, and restoration; network management; network security; network design, measurements, and modeling; broadband access; MSO/cable networks; VoIP and IPTV; transmission media and systems; switching and routing (from legacy to next-generation paradigms); telecommunication software; wireless communication systems; wireless, cellular and personal networks; satellite and space communications and networks; optical communications and networks; free-space optical communications; cognitive communications and networks; green communications and networks; heterogeneous networks; dynamic networks; storage networks; ad hoc and sensor networks; social networks; software defined networks; interactive and multimedia communications and networks; network applications and services; e-health; e-business; big data; Internet of things; telecom economics and business; telecom regulation and standardization; and telecommunication labs of all kinds. Proposals of interest should suggest textbooks that can be used to design university courses, either in full or in part. They should focus on recent advances in the field while capturing legacy principles that are necessary for students to understand the bases of the discipline and appreciate its evolution trends. Books in this series will provide high-quality illustrations, examples, problems and case studies.

For further information, please contact: Dr. Tarek S. El-Bawab, Series Editor, Professor and Dean of Engineering, American University of Nigeria, tel-bawab@ieee.org; or Mary James, Senior Editor, Springer, mary.james@springer.com

More information about this series at http://www.springer.com/series/13835

Tetsuya Kawanishi

Electro-optic Modulation for Photonic Networks

Precise and high-speed control of lightwaves

 Springer

Tetsuya Kawanishi
Department of Electronic and Physical
Systems
Waseda University
Tokyo, Japan

ISSN 2524-4345   ISSN 2524-4353   (electronic)
Textbooks in Telecommunication Engineering
ISBN 978-3-030-86722-5   ISBN 978-3-030-86720-1   (eBook)
https://doi.org/10.1007/978-3-030-86720-1

This Springer imprint is published by the registered company Springer Nature Switzerland AG
The registered company address is: Gewerbestrasse 11, 6330 Cham, Switzerland

Preface

This text book is on optical modulators using electro-optic (EO) effect which controls the optical phase. The optical modulators consist of electrodes for high-speed electric signals and optical waveguides for lightwaves, where the electric fields induced by the electric signals interact with the lightwaves. Thus, we should have knowledge of high-frequency electronics and optical waveguide devices to understand operation principles of high-speed optical modulators. This book offers overviews of various types of optical modulators, but it focuses on lithium niobate (LN) modulators which show ideal EO effect even in high-frequency region. The operation of the LN modulator can be precisely described by the mathematical model based on Bessel functions. The LN modulators are commonly used in commercial optical fiber transmission systems, optical measurement instruments, etc. In addition, the knowledge on the LN modulator is useful to learn other types of modulators such as semiconductor-based optical modulators.

The optical modulators, bridging electric and optical signals, are at the boundary between physics and systems, where digital information is converted into lightwaves. This book provides knowledge both on optical devices and modulation formats, which is required to design systems using optical modulators. I wish to bridge the gap between physics and systems, by concise mathematical expressions and physical models. For example, this book uses a unique simple expression for the EO effect, to discuss the operation of the modulators, while the effect is commonly described by a refractive index ellopsoid in conventional textbooks. The modulator output can be precisely expressed by the Bessel function; thus, this book offers some mathematical expressions based on the Bessel function.

I gratefully acknowledge a debt to my colleagues and friends, especially to Dr. Atsushi Kanno, Dr. Pham Tien Dat, Dr. Keizo Inagaki, and Dr. Yuya Yamaguchi of National Institute of Information and Communications Technology (NICT), Japan, for their generous cooperation. I would like to thank Dr. Yoshihiro Ogiso of NTT Device Innovation Center, Nippon Telegraph and Telephone Corporation, Japan. He kindly provided measured data and figures on semiconductor modulators. Some research work described in this book was carried out in collaboration with Sumitomo Osaka Cement (SOC). I would like to thank Mr. Junichiro Ichikawa of SOC for his support. My thanks are also due to the students in my group at Waseda University, especially to Mr. Zu-Kai Weng and Mr. Yu Zhou. Finally, I would like to express my sincere appreciation to my wife, Noriko, and my daughter, Yuriko, for their enormous patience and steady support.

Tokyo, Japan   Tetsuya Kawanishi

Contents

Chapter 1
Introduction

Optical modulation plays important roles in photonic networks, including optical fiber links, visible ray communications, and fiber wireless, as well as in high-performance sensing systems [1, 2]. The aim of this book is to provide comprehensive and detailed information on electro-optic modulation that offers high-speed and precise control of lightwaves. While this book will review various types of optical modulation devices, it will focus on modulators based on electro-optic (EO) effect (Pockels effect), which are called electro-optic (EO) modulators, where optical phase retardation is induced by electric field applied thorough electrodes [3–5]. Various types of functions, including phase modulation, intensity modulation, and vector modulation, can be achieved by the use of optical interferometers with the EO effect. Mach–Zehnder interferometer modulators (MZMs) are very useful for the most common modulation formats such as on–off-keying (OOK) and binary-phase-shift-keying (BPSK). In addition, MZMs would be elements for more complicated modulators, high-speed optical switches, vector modulators, etc.

Particular materials, such as lithium niobate (LN) [6, 7] and lithium tantalate (LT) [8], provide linear optical phase shift that is in proportion to intensity of the electric field. Optical waveguides formed in LN substrates offer low-loss and stable optical signal propagation inside modulators. The LN modulators are commonly used for high-end applications such as long-haul fiber communications.

This book provides mathematical expressions that are needed to describe the basic principles and particular functions of EO modulators. Optical modulation in LN is almost exactly expressed by a series of Bessel functions. The difference in modulator characteristics between the mathematical expressions and measured results by experiments would be negligibly small. That means that the mathematical expressions can precisely describe LN EO modulator operation and would be a good example of applied mathematics in optical communication technologies. The mathematical expressions would be also useful for other modulators besides LN modulators. For example, semiconductor MZMs using optical phase modulation can be approximately described by the mathematical expressions in this book, although they would have some parasitic intensity excursion.

© Springer Nature Switzerland AG 2022
T. Kawanishi, *Electro-optic Modulation for Photonic Networks*, Textbooks in
Telecommunication Engineering, https://doi.org/10.1007/978-3-030-86720-1_1

This book has eight chapters, including this chapter for introduction. As described earlier, optical modulation and modulators play important roles in modern photonic networks where precise and high-speed lightwave control is required for high bit-rate optical transmission or high-performance analog waveform transfer. In this book, operation principles of EO modulators are shown by concise mathematical expressions that help in the understanding of sideband generation. First, basic functions of optical modulators for various modulation schemes are described by time domain descriptions. Second, spectral components generated by modulators are clearly expressed by Bessel functions. This book also provides practical knowledge of optical modulators as well as basic theory on modulator operation. It stresses the importance of optical phase fluctuation. Understanding of optical phase stabilization and control would be very important to design actual optical communication systems.

Chapter 2 illustrates the roles of modulators in photonic networks. As described earlier, this book will focus on high-performance EO modulators that can be used in optical telecommunication systems such as long-haul large capacity optical fiber links [9–11], free-space optic links [12, 13] and fiber-radio systems [14, 15]. These systems would be indispensable items in future networks where radio and optical technologies are integrated together. The operation principle of the modulator would help in the understanding of basics of optical switching also. Such information and knowledge will be useful to consider ultimate limitation of telecommunication systems due to fundamental physical issues such as upper limit of light propagation speed, etc. This chapter also discusses the basics of modulation and bandwidths of modulated signals. The number of available channels in an optical fiber link can be calculated from the bandwidths.

Chapter 3 shows various modulation techniques, including direct modulation [16] and external modulation [17]. The direct modulation, where an optical output power can be controlled by an electric current injected into a laser, can offer a simple optical signal source; however, it causes large parasitic phase excursion that broadens the signal bandwidth [18]. In the external modulation, a static lightwave emitted from a laser is modulated by an external modulator [3–5]. While various phenomena can be used for external modulators, this book focuses on EO modulators using ferroelectric materials, such as LN and LT, which offer ideal phase and amplitude modulation.

Chapter 4 describes the basics of modulators based on EO effect, by using time domain mathematical expressions. First, we show an operation principle of phase modulators, by which various types of modulators can be constructed. An MZM consisting of two phase modulators offers intensity modulation, with push–pull operation where the induced optical phases at the two phase modulators have the opposite signs to each other. Vector modulation can be provided by an integrated modulator with two MZMs, which modulate in-phase and quadrature components independently. This chapter also describes various device structures with transmission lines on LN substrates for high-speed operation, and configurations of electric circuits for push–pull operation. In the last part of this chapter, we discuss the stability issue caused by optical phase fluctuation due to laser noise and mechanical vibration. Integrated MZMs with automatic bias control can offer stable intensity or amplitude modulation.

Chapter 5 describes configurations of modulators for digital signal transmission with various modulation formats. In principle, we can consider any configurations that offer arbitrary signal generation, but taking into account noise or fluctuation in actual electrical and optical signals, only a limited number of configurations are practically advantageous. First, this chapter describes binary modulation formats that are commonly used for digital communication systems. An MZM can be used for phase-shift-keying (PSK) an well as for on–off-keying (OOK). Multilevel modulation formats are also reviewed, where multilevel signals are synthesized by electric or optical circuits.

Chapter 6 provides frequency domain expressions on sideband components generated in an EO modulator, to discuss the behavior of optical modulators for high-speed signals. Actual signals with information would have many frequency components; however, we can understand the operation principles of EO modulators by considering sideband components generated from a sinusoidal wave. Optical modulation with a sinusoidal signal has many applications in analog transmission [2], radio astronomy [19], and device measurement [20–22]. The Bessel function can offer a concise expression for a series of sidebands generated by phase modulation with sinusoidal signals. Operation principles of various modulators based on phase modulation also can be expressed by frequency domain mathematical expressions with the Bessel function. In an MZM, the output spectrum depends on bias conditions and also on imbalance of the modulator structure.

Chapter 7 describes modulator configuration for double sideband (DSB) [3, 20] and single sideband (SSB) [23, 24]. DSB signals have symmetric spectra with respect to the carrier, while SSB signal spectra are not symmetric. In DSB modulation, there is no phase difference between signals on two phase modulators for push-pull operation. On the contrary, in SSB modulation, USB or LSB can be suppressed largely by controlling the phase difference between the signals. In this chapter, we focus on the carrier and the first order USB and LSB, which are commonly used as desired components in actual systems.

Chapter 8 provides various methods for estimation of optical modulator parameters. The mathematical expressions defined in Chap. 6 precisely describe output spectra of actual LN optical modulators, where refractive index change in LN is precisely proportional to the voltage applied to the modulator. Thus, various types of parameters, which describe performance of the modulators, can be precisely estimated from ratios of sideband components [25, 26].

References

1. T. Kawanishi, *Wired and Wireless Seamless Access Systems for Public Infrastructure* (Artech House, London, 2020)
2. T. Kawanishi, A. Kanno, H.S.C. Freire, Wired and wireless links to bridge networks: seamlessly connecting radio and optical technologies for 5G networks. IEEE Microw. Mag. **19**(3), 102–111 (2018)

3. T. Kawanishi, T. Sakamoto, M. Izutsu, High-speed control of lightwave amplitude, phase, and frequency by use of electrooptic effect. IEEE J. Select. Top. Quantum Electron. **13**(1), 79–91 (2007)

4. T. Kawanishi, Integrated Mach-Zehnder interferometer-based modulators for advanced modulation formats, in *High Spectral Density Optical Communication Technologies, Optical and Fiber Communications Reports, vol. 6*, ed. by M. Nakazawa, K. Kikuchi, T.Miyazaki (Wiley, New York, 2010)

5. T. Kawanishi, High-speed optical communications using advanced modulation formats, in *Wiley Encyclopedia of Electrical and Electronics Engineering*, ed. by J.G. Webster (Wiley, New York, 2016)

6. R.C. Alferness, Waveguide electrooptic modulators. IEEE Trans. Microw. Theory Tech. **30**(8), 1121–1137 (1982)

7. M. Izutsu, Y. Yamane, T. Sueta, Broad-band traveling-wave modulator using a LiNbO$_3$ optical waveguide. IEEE J. Quantum Electron. **13**(4), 287–290 (1977)

8. H. Murata, K. Kinoshita, G. Miyaji, A. Morimoto, T. Kobayashi, Quasi-velocity-matched LiTaO$_3$ guided-wave optical phase modulator for integrated ultrashort optical pulse generators. Electron. Lett. **36**(17), 1459–1460 (2000)

9. A. Sano, H. Masuda, Y. Kisaka, S. Aisawa, E. Yoshida, Y. Miyamoto, M. Koga, K. Hagimoto, T. Yamada, T. Furuta, H. Fukuyama, 14-Tb/s (140x111-Gb/s PDM/WDM) CSRZ-DQPSK transmission over 160km using 7-THz bandwidth extended L-band EDFAs, in *ECOC 2006 Proceedings*, vol. 6, 2006

10. A.H. Gnauck, G. Charlet, P. Tran, P.J. Winzer, C.R. Doerr, J.C. Centanni, E.C. Burrows, T. Kawanishi, T. Sakamoto, K. Higuma, 25.6-Tb/s WDM transmission of polarization-multiplexed RZ-DQPSK signals. J. Lightwave Technol. **26**(1), 79–84 (2008)

11. P.J. Winzer, A.H. Gnauck, 112-Gb/s polarization-multiplexed 16-QAM on a 25-GHz WDM grid, in *2008 34th European Conference on Optical Communication* (IEEE, New York, 2008), pp. 1–2

12. T. Umezawa, T. Sakamoto, A. Kanno, A. Matsumoto, K. Akahane, N. Yamamoto, T. Kawanishi, 25-Gbaud 4-WDM free-space optical communication using high-speed 2-D photodetector array. J. Lightw. Technol. **37**(2), 612–618 (2019)

13. T. Umezawa, Y Yoshida, A. Kanno, A. Matsumoto, K. Akahane, N.Yamamoto, T. Kawanishi, FSO receiver with high optical alignment robustness using high-speed 2D-PDA and space diversity technique. J. Lightw. Technol. **39**(4), 1040–1047 (2021)

14. A. Kanno, K. Inagaki, I. Morohashi, T. Sakamoto, T. Kuri, I. Hosako, T. Kawanishi, Y. Yoshida, K. Kitayama, 40 Gb/s W-band (75-110 GHz) 16-QAM radio-over-fiber signal generation and its wireless transmission. Opt. Express **19**(26), B56–B63 (2011)

15. P.T. Dat, Y. Yamaguchi, K. Inagaki, M. Motoya, S. Oikawa, J. Ichikawa, A. Kanno, N. Yamamoto, T. Kawanishi, Transparent fiber-radio-fiber bridge at 101 GHz using optical modulator and direct photonic down-conversion, in *Optical Fiber Communication Conference*, p. F3C.4 (Optical Society of America, San Diego, 2021)

16. Y. Matsui, D. Mahgerefteh, X. Zheng, C. Liao, Z.F. Fan, K. McCallion, P. Tayebati, Chirp-managed directly modulated laser (CML). IEEE Photon. Technol. Lett. **18**(2), 385–387 (2006)

17. G.L. Li, P.K.L. Yu, Optical intensity modulators for digital and analog applications. J. Lightw. Technol. **21**(9), 2010–2030 (2003)

18. F. Koyama, K. Iga, Frequency chirping in external modulators. J. Lightw. Technol. **6**(1), 87–93 (1988)

19. H. Kiuchi, T. Kawanishi, M. Yamada, T. Sakamoto, M. Tsuchiya, J. Amagai, M. Izutsu, High extinction ratio Mach-Zehnder modulator applied to a highly stable optical signal generator. IEEE Trans. Microw. Theory Tech. **55**(9), 1964–1972 (2007)

20. K. Inagaki, T. Kawanishi, M. Izutsu, Optoelectronic frequency response measurement of photodiodes by using high-extinction ratio optical modulator. IEICE Electron. Express **9**(4), 220–226 (2012)

21. Measurement method of a half-wavelength voltage and a chirp parameter for Mach-Zehnder optical modulator in high-frequency radio on fibre (RoF) systems, IEC 62802:2017

22. Transmitting equipment for radiocommunication - Frequency response of optical-to-electric conversion device in high-frequency radio over fibre systems - Measurement method, IEC 62803:2016

23. M. Izutsu, S. Shikamura, T. Sueta, Integrated optical SSB modulator/frequency shifter. IEEE J. Quant. Electron. **17**(11), 2225–2227 (1981)

24. T. Kawanishi, M. Izutsu, Linear single-sideband modulation for high-SNR wavelength conversion. IEEE Photon. Technol. Lett. **16**(6), 1534–1536 (2004)

25. T. Kawanishi, K. Kogo, S. Oikawa, M. Izutsu, Direct measurement of chirp parameters of high-speed Mach-Zehnder-type optical modulators. Opt. commun. **195**(5–6), 399–404 (2001)

26. S. Oikawa, T. Kawanishi, M. Izutsu, Measurement of chirp parameters and halfwave voltages of Mach-Zehnder-type optical modulators by using a small signal operation. IEEE Photon. Technol. Lett. **15**(5), 682–684 (2003)

Chapter 2
Role of Optical Modulation in Photonic Networks

For photonic networks, optical modulation is widely used to generate optical signals in transmitters. Optical modulators modulate one or more parameters, such as amplitude, phase, and frequency, which describes waveforms of lights. Modulators and modulation technologies were originally developed for electric signals. Basic principles of optical modulation are the same as in electric modulation. However, device structures, materials, and principles would have significant differences between electric and optical modulators. Electric signals are converted into optical signals by optical modulators embedded in optical transmitters. This chapter describes roles of optical modulators in optical transmitters, optical measurement systems, etc. Basics of optical modulation are also provided to discuss bandwidths of optical signals. The number of available optical channels in a fiber link can be calculated from the bandwidths.

2.1 Optical Modulators for Transmission

Photonic networks consist of various optical transmission systems such as intercontinental long-haul links, inter-city metro links, and access links connecting end-users. As shown in Fig. 2.1, optical transmission systems comprise optical transmitters, optical fibers, and optical receivers. Single-mode fibers (SMFs) offer low-loss transmission for broadband signals. Optical transmission systems should have electric-to-optical (EO) signal conversion devices at transmitters, and optical-to-electric (OE) devices at receivers, because digital information is generated and processed by electric circuits in user equipment, sensors, switches, and servers. Optical modulators in the transmitters convert electric signals into optical signals for optical fiber transmission, while photodetectors convert the optical signals into electric signals in the receivers [1–3].

Optical signals carrying information is generated by the EO signal conversion devices which control amplitude, phase, or frequency of lightwaves. This function is

© Springer Nature Switzerland AG 2022
T. Kawanishi, *Electro-optic Modulation for Photonic Networks*, Textbooks in
Telecommunication Engineering, https://doi.org/10.1007/978-3-030-86720-1_2

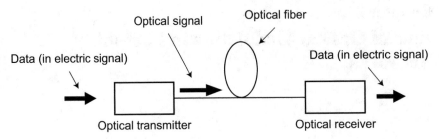

Fig. 2.1 Basic configuration of optical fiber transmission

called modulation. On the other hand, the OE signal conversion devices extract the information from the optical signals. This is called demodulation.

Figure 2.2 shows an example of optical modulation by an optical modulator, where lightwave amplitude is controlled by an electric signal. The amplitude of the optical output is proportional to the voltage of the electric signal. This modulation scheme is called amplitude modulation. For demodulation, a photodetector can convert envelope of the amplitude modulation signal into an electric signal directly. Frequency and phase modulation can also be used for photonic networks. However, frequency or phase shift in optical signals should be converted into optical amplitude or intensity change for demodulation. In principle, photodetectors are not sensitive to optical frequency or phase. They can only detect envelopes of optical signals.

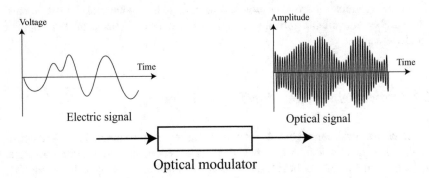

Fig. 2.2 Optical amplitude modulation

In a digital transmission system, a bit stream is carried by symbols mapped to the states of the lightwave. Figure 2.3 shows a time domain profile of a binary modulation signal. As described above, amplitude modulation is commonly used in actual transmission systems. Modulation in frequency or phase can be used for data transmission. The modulation speed is commonly described in baud rate, which is defined by the number of symbol changes per second, as follows:

$$f_\mathrm{p} = \frac{1}{T_\mathrm{p}}, \tag{2.1}$$

where T_p is the time duration of the symbols. The transmission capacity is equal to the baud rate, when the optical signal is modulated by a simple modulation format using two symbols mapped into 0 and 1 [1].

Figure 2.4 shows a time domain profile of a multilevel modulation signal. By using multilevel modulation formats with many symbols, the total transmission capacity can be increased. As discussed in Sect. 2.2.3, lightwaves can be described by phasors, as well as by wave parameters: amplitude, phase, and frequency. The phasors can be expressed by two-dimensional vectors. Thus, we can increase the number of symbols by modulating each component in the vectors. This is called vector modulation, where but precise lightwave control is required to increase the number of symbols. The transmission capacity, i.e., bit rate (bits per second), can be expressed by

$$C = f_p \log_2 N_p, \tag{2.2}$$

where N_p is the number of the symbols used in the modulation formats [1, 4, 5].

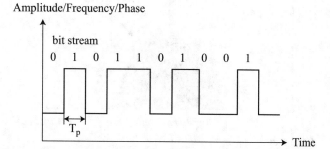

Fig. 2.3 Binary modulation signal

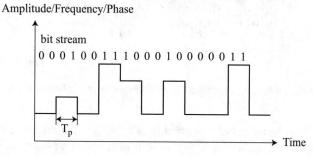

Fig. 2.4 Multilevel modulation signal

Figure 2.5 shows frequency and time domain profiles of a modulated signal, which has a finite bandwidth. The bandwidth of the modulated signal depends on the baud rate. For modulation with an analog signal, the bandwidth is described by the frequency of the modulating signal. As an example, we consider amplitude modulation by a sinusoidal wave signal with the configuration shown in Fig. 2.2. The bandwidth of the optical signal would be expressed by the frequency of the electric signal. Section 2.2 offers a simple mathematical model for bandwidths of modulated signals. More details will be given in Chap. 6.

To increase the bit rate with a limited bandwidth, we should increase the number of the symbols. Bit rate per unit bandwidth (1 Hz), which is called spectral efficiency, can be enhanced by modulation formats with many symbols. However, precise control of lightwave at the modulator is required to generate optical signals with many symbols.

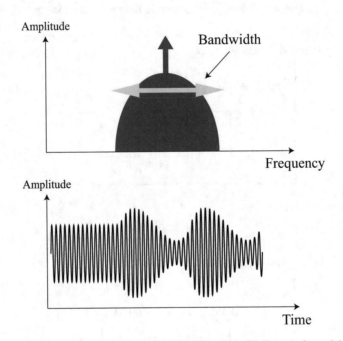

Fig. 2.5 Frequency domain spectrum (upper) and time domain profile (lower) of a modulated signal

Figure 2.6 shows a schematic of a multiplexing system which divides the resources of the transmission media into channels, in wavelength (frequency), time, or space domain. Such multiplexing systems can provide many optical channels to increase the total transmission capacity. As described in Sect. 2.2.4, wavelength-division multiplexing (WDM) is commonly used in commercial transmission systems, where the total transmission capacity can be larger than 10 Tb/s [6, 7]. The number of channels can be given by the ratio of the bandwidth of the signal to the total available bandwidth.

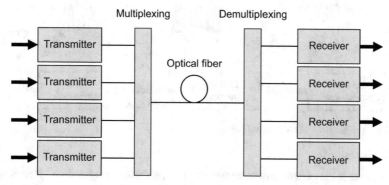

Fig. 2.6 Transmission system with multiplexing

In conventional optical transmission systems, the available optical frequency bandwidth was large enough for data traffic demands, so that simple modulation formats have been commonly used, where enhancement of spectral efficiency was not an important issue. Thus, more effort was put on high-speed operation rather than multilevel modulation formats for high spectral efficiency. However, now, the bandwidth provided by SMFs is not large enough to meet the demands of data transmission. Thus, optical transmission systems using multilevel modulation formats have been developed recently, to achieve enhanced spectral efficiency. In general, there is a trade-off between modulation speed and preciseness [8]. However, to realize huge capacity optical transmission whose capacity is over Tb/s, we should pursue precise modulation for modulation formats with many symbols and high-speed modulation for high baud rate signal generation, simultaneously. To increase the number of symbols effectively, we can use high-speed vector modulation using integrated external optical modulators, where the sine and cosine components are controlled independently [9–14].

In wired and wireless seamless networks, analog optical transmission systems, which transfer waveforms for radio services over fibers, can be used for photonic networks connecting many base stations for mobile services [1, 8, 15–18]. Figure 2.7 shows a configuration of a radio-over-fiber (RoF) system, as an example of analog transmission systems. The waveform of the radio wave signal is converted into an optical signal whose envelop is proportional to the waveform. At the receiver side, the radio wave retrieved by a photodetector can be transmitted through an antenna. In RoF, slight imperfection in optical modulation would have significant impact on the quality of the radio wave for emission, where spurious components should be largely suppressed to mitigate undesired interference [1].

As described above, high-speed and precise modulation is indispensable for high-speed transmission with multilevel modulation formats and analog transmission for radio services [1, 15, 17, 19–21]. In addition, precise characterization of optical

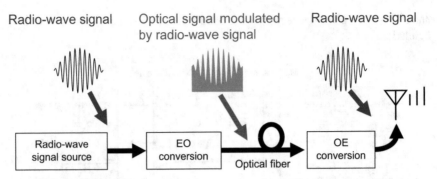

Fig. 2.7 Configuration of radio over fiber

components is very important to construct high-performance photonic systems with multilevel modulation or analog transmission [1, 22–24]. In optical modulation, frequency shifted spectral components can be generated by feeding a sinusoidal electric signal to a modulator, as shown in Fig. 2.8. These spectral components are called sidebands, where the frequency separation between sidebands is precisely equal to the frequency of the electric signal. Section 2.2 and Chap. 6 provide more details on sideband generation in optical modulation. A part of sidebands generated by the modulator can be used as standard signals or stimulus signals for optoelectronic device measurement [25]. We can find more details in Sect. 8.3. Undesired sidebands can be largely suppressed by using precise optical modulation.

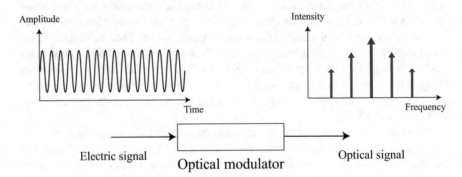

Fig. 2.8 Optical sideband generation by modulation

2.2 Signal Bandwidth and Modulation

Modulation broadens spectra of lightwaves. Spectral width difference between input and output lightwaves of an optical modulator is proportional to a bandwidth of information which would be transferred by the modulator. This chapter describes such spectral broadening by a simple mathematical model.

2.2.1 Basic Configuration of Optical Modulation

An electric field of a monochromatic lightwave can be described by a sinusoidal function as follows:

$$E_{in}(t) = A_0 \cos(\omega_0 t + \phi_0), \tag{2.3}$$

which is called an unmodulated lightwave or signal. A_0, ω_0, and ϕ_0 are, respectively, the amplitude, angular frequency, and phase of the electric field. In this section, we use this real function expression, for simplicity, while the lightwave can also be described by a complex number with phasor expression as in Sect. 4.1.

Figure 2.9 shows a basic configuration of optical modulation. An input lightwave (unmodulated signal E_{in}) and an electric signal (modulation signal whose waveform is $f(t)$) are fed to an optical modulator to induce an external electric field. While various types of physical phenomena through electric field, electric current, electric power, etc. can also be used for optical modulation, this book focuses on optical modulation by electro-optic effect which generates interaction between the lightwave in the modulator and the external electric field. The modulator outputs a modulated signal described by E_{out}.

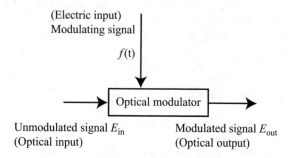

Fig. 2.9 Optical modulation

Modulation is the function of varying one or more properties including amplitude, phase, and frequency of a periodic waveform. Modulation with amplitude is called amplitude modulation (AM). Phase modulation (PM) and frequency modulation (FM) are through varying phase and frequency, respectively. Such modulation techniques can be combined to generate complicated waveforms.

The modulated signal E_{out} is given by

$$E_{out}(t) = A(t) \cos\left[\omega(t)t + \phi(t)\right], \tag{2.4}$$

where $A(t)$, $\phi(t)$, and $\omega(t)$ are time varying amplitude, phase, and angular frequency, respectively. The frequency can be given by $\omega(t)/2\pi$.

The phase and frequency (angular frequency) comprise the argument of the cosine function in (2.4). In other words, PM and FM vary the angle of the triangular function, so that they are called angle modulation. The following sections describe change of spectra by amplitude and angle modulation.

2.2.2 Amplitude Modulation

Amplitude modulation (AM) is one of the simplest modulation formats, where phase and frequency are constant. The time varying amplitude $A(t)$ is expressed by

$$A(t) = 1 + f(t), \tag{2.5}$$

where $f(t)$ is an arbitrary real function. As a simple example, we consider AM with a cosine function described by

$$f(t) = M \cos \omega_m t, \tag{2.6}$$

where $f_m = \omega_m/2\pi$ and M are called modulation frequency and modulation index, respectively.

Assuming phase and frequency are constant, they can be described by $\omega(t) = \omega_0$ and $\phi(t) = \phi_0$. By defining the origin of the temporal axis properly, the constant phase ϕ_0 can be assumed to be 0, without loss of generality.

The amplitude $A(t)$ is given by

$$A(t) = 1 + M \cos \omega_m t, \tag{2.7}$$

so that the modulated signal can be expressed by

$$E_{\text{out}}(t) = (1 + M \cos \omega_m t) \cos \omega_0 t. \tag{2.8}$$

By using a product-to-sum trigonometric formula,

$$\cos \alpha \cos \beta = \frac{1}{2} \left[\cos(\alpha + \beta) + \cos(\alpha - \beta) \right], \tag{2.9}$$

the modulated signal can be described by

$$E_{\text{out}}(t) = \cos \omega_0 t + \frac{M}{2} \left[\cos(\omega_0 + \omega_m)t + \cos(\omega_0 - \omega_m)t \right], \tag{2.10}$$

where two spectral components whose angular frequencies are $\omega_0 + \omega_m$ are $\omega_0 - \omega_m$ generated besides the input lightwave component of ω_0. In other words, AM generates an optical signal with three spectral components from an optical input with one spectral component.

The two new spectral components are called sidebands or sideband components, while the spectral component of the input lightwave frequency is called carrier. The sidebands whose frequencies are lower and higher than that of the carrier are called lower sideband (LSB) and upper sideband (USB), respectively. The frequencies of the carrier, USB, and LSB are, respectively, $f_0, f_0 + f_m$, and $f_0 - f_m$, where angular frequencies can be converted into frequencies by $2\pi f_0 = \omega_0$ and $2\pi f_m = \omega_m$.

Equation (2.10) shows that an unmodulated lightwave with a single frequency f_0 into a modulated lightwave whose spectral components are in a frequency region from $f_0 - f_m$ to $f_0 + f_m$. The bandwidth and center frequency of the modulated signal are, respectively, f_0 and $2f_m$, which means that modulation broadens optical spectral width. The bandwidth of the modulated signal would be double the bandwidth of the modulating signal which carries information, where the bandwidth of the unmodulated lightwave is almost zero, if we use a narrow linewidth laser.

To discuss modulation with complicated signals, we consider a modulating signal with multiple spectral components, described by

$$f(t) = \sum_{N=1}^{N_m} M_N \cos\left(\omega_{m_N} t + \phi_N\right), \qquad (2.11)$$

where N_m is the number of the spectral components.

The amplitude is expressed by

$$A(t) = 1 + \sum_{N=1}^{N_m} M_N \cos\left(\omega_{m_N} t + \phi_N\right), \qquad (2.12)$$

and then the modulated signal is given by

$$E_{\text{out}}(t) = \cos\omega_0 t + \sum_{N=1}^{N_m} \frac{M_N}{2} \left[\left\{\cos(\omega_0 + \omega_{m_N})t + \phi_N\right\}\right.$$
$$\left. + \left\{\cos(\omega_0 - \omega_{m_N})t - \phi_N\right\}\right]. \qquad (2.13)$$

This equation shows that modulation with a multiple spectral signal generates a USB and an LSB from each spectral component in the modulating signal. The modulated signal would be in the range from $f_0 - f_{N_m}$ to $f_0 + f_{N_m}$, where $f_{N_m} = \omega_{N_m}/2\pi$ is the largest frequency of the spectral components in the modulating signal.

A modulating signal whose fundamental frequency is $f_m (= \omega_m/2\pi)$ can be expressed by the Fourier expansion as follows:

$$f(t) = \sum_{N=1}^{\infty} M_N \cos(N\omega_m t + \phi_N). \qquad (2.14)$$

In general, amplitudes of high-order terms would be very small. In addition, actual electric circuits have upper limits in their frequency responses, so that we can neglect high-order terms. When the frequency upper limit is $N_m f_m$, (2.14) would be identical to (2.11), by assuming $\omega_{m_{N_m}} = N_m \omega_m$.

To obtain a more general expression, we replace the first term in (2.12) by $M_0 = 1$ and assume $\omega_{m_0} = \phi_0 = 0$. The amplitude can be expressed by

$$A(t) = \sum_{N=0}^{N_m} M_N \cos\left(\omega_{m_N} t + \phi_i\right). \tag{2.15}$$

2.2.3 Angle Modulation

This subsection describes angle modulation which includes FM with $\omega(t)$ and PM with $\phi(t)$ in (2.4). The phase at a particular moment $\Phi(t)$, which is called instantaneous phase, can be defined by phase excursion from unmodulated oscillation with angular frequency of ω_0, as follows:

$$\Phi(t) \equiv [\omega(t) - \omega_0]t + \phi(t). \tag{2.16}$$

By using this instantaneous phase, (2.4) can be rewritten by

$$E_{\text{out}}(t) = A(t) \cos\left[\omega_0 t + \Phi(t)\right], \tag{2.17}$$

where FM and PM can be comprehensively expressed.

Frequency excursion from the carrier frequency f_0 can be described by derivative of the instantaneous phase, as follows:

$$\Delta f = f - f_0$$
$$= \frac{1}{2\pi} \frac{d\Phi(t)}{dt}, \tag{2.18}$$

where f describes a frequency at a particular time and is called an instantaneous frequency. Figure 2.10 shows relationship between the frequency excursion and the instantaneous phase. For example, if Φ continuously increases with a constant differential coefficient, the frequency excursion Δf is a positive constant. The instantaneous frequency $f(t)$ would be a constant and higher than the carrier frequency.

By using the trigonometric addition theorem, (2.17) can be rewritten as

$$E_{\text{out}}(t) = A(t) \cos \Phi(t) \cos(\omega_0 t) - A(t) \sin \Phi(t) \sin(\omega_0 t)$$
$$= A_{\text{I}}(t) \cos(\omega_0 t) - A_{\text{Q}}(t) \sin(\omega_0 t), \tag{2.19}$$

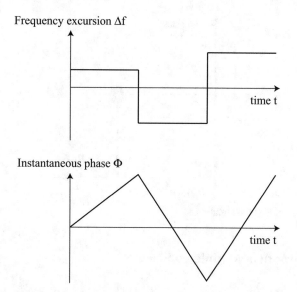

Fig. 2.10 Relationship between PM and FM

where $A_I(t)$ and $A_Q(t)$ are defined by

$$A_I(t) \equiv A(t) \cos \Phi(t) \tag{2.20}$$

$$A_Q(t) \equiv A(t) \sin \Phi(t). \tag{2.21}$$

Based on discussion on sideband generation from $A(t)$ with AM described in (2.15), we can deduce that $A_I(t)$ and $A_Q(t)$ would generate USB and LSB to broaden the spectral width.

Figure 2.11 shows a phasor with a reference of $\cos(\omega_0 t)$ corresponding to the unmodulated input lightwave, where $A_I(t)$ and $A_Q(t)$ are described as a real and an imaginary components. The imaginary component, which is proportional to $-\sin(\omega_0 t)$, has a $\pi/2$ phase lead with respect to the real component, which is proportional to $\cos(\omega_0 t)$. These two components are time varying, so that the state of the lightwave described by the phasor would move over the two-dimensional complex plane.

$A_I(t)$, which has the same phase with $\cos(\omega_0 t)$, is called I- or in-phase component, while $A_I(t)$ is called Q- or quadrature component.

Mathematical expressions with Bessel's functions are needed for more rigorous explanations of sideband generation in angular modulation. Details will be provided in Chap. 6.

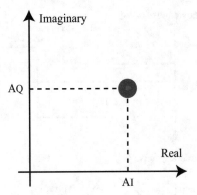

Fig. 2.11 Phasor of angle modulation

2.2.4 Bandwidth of Modulated Signals

As we discussed in the previous subsections, amplitude and angle modulation generate sideband components (USB and LSB) whose frequency separation from the carrier is equal to the frequency of the modulating signal. When the bandwidth of the (electric) modulating signal is f_{BW}, the (optical) modulated signal would have a bandwidth given by

$$\Delta f = 2 f_{BW}. \tag{2.22}$$

Periods of electric signals are often described by frequencies, while wavelengths are often used to describe periods or frequencies of optical signal waveforms. In this book, we use wavelengths in vacuum. While various frequency components are generated from random bit sequences, the highest frequency component can be associated with the bit pattern of "01010101 . . .." As shown in Fig. 2.12, the bit pattern can be approximated by a sinusoidal waveform whose period of the sinusoidal waveform is T_k, where $T_k = 2T_p$. The frequency of this waveform can be given by

$$
\begin{aligned}
f_k &= \frac{1}{T_k} \\
&= \frac{1}{2T_k} = \frac{1}{2} f_p, \tag{2.23}
\end{aligned}
$$

so that the highest frequency component in the random bit sequences is equal to a half of the baud rate (f_p). By approximating the bandwidth of the modulating signal by f_k, the signal bandwidth can be expressed by

$$\Delta f = 2 f_k = f_p. \tag{2.24}$$

Thus, the bandwidth of the electric modulating signal is equal to the baud rate. As mentioned above, the bandwidth of the modulating signal is a half of the baud rate, because the bit pattern of "01010101" can be described by a sinusoidal wave whose frequency is equal to a half of the baud rate. The optical modulator generates two sidebands: USB and LSB, so that the total bandwidth is double the bandwidth of the electric modulating signal.

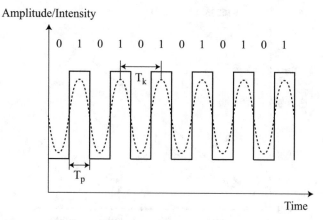

Fig. 2.12 Bit pattern in which "01" repeats

By using the relationship between wavelength λ and frequency f, $f = c/\lambda$, the bandwidth of frequency domain can be converted to that of wavelength domain $\Delta\lambda$, as follows:

$$\Delta\lambda = \left| \left(\frac{\mathrm{d}f}{\mathrm{d}\lambda} \right)^{-1} \right| \Delta f$$

$$= \frac{\lambda^2}{c} \Delta f, \tag{2.25}$$

where c is the speed of light, and

$$\frac{\mathrm{d}f}{\mathrm{d}\lambda} = -\frac{c}{\lambda^2}. \tag{2.26}$$

Here, we assumed that $\Delta\lambda \ll \lambda$.

In optical fiber systems, lightwaves whose wavelengths are close to 1310 and 1550 nm are commonly used, because loss in single mode fibers is lowest at 1550 nm, and dispersion is smallest at 1310 nm. In optical fiber communication systems, f_k is smaller than 100 GHz, so that $\Delta\lambda$ is much smaller than λ. Frequency domain bandwidth Δf in GHz can be converted to wavelength domain bandwidth $\Delta\lambda$ in nm, by

$$\Delta\lambda = 3.3 \times 10^{-9} \lambda^2 \Delta f, \tag{2.27}$$

where curves for $\Delta\lambda$ as functions of Δf are shown in Fig. 2.13. The center wavelength of the signal λ is also in nm. The wavelengths are 1550, 1310, and 633 nm. For example, $\Delta\lambda$ at 1550 nm would be 0.8nm when Δf is equal to 100 GHz (f_k = 50 GHz).

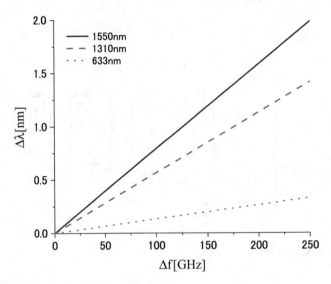

Fig. 2.13 Bandwidths in frequency and wavelength

2.3 Wavelength-Division Multiplexing

WDM has been deployed in long-haul and large-capacity optical fiber transmission systems to increase the number of available optical channels in a fiber [6, 7]. In a WDM transmission system, a multiplexing device combines optical signals generated by transmitters, where the channels are allocated with equal separation in optical frequency domain. Each transmitter emits an optical signal at an assigned wavelength (or frequency), where λ_1, λ_2, λ_3, and λ_4 denote the first, the second, the third, and the fourth WDM channels, as shown in Fig. 2.14. At the receiver side, each WDM channel is extracted by a demultiplexing device. Optical filters can be used for the multiplexing and demultiplexing.

Wavelength region close to 1550 nm where loss is lowest is commonly used for such WDM systems. The optical frequency of the lightwave whose wavelength is 1550 nm is, approximately, 193 THz. Optical frequencies of WDM channels are defined by an ITU (International Telecommunication Union) recommendation (ITU-T G694.1) [26]. Frequency allocation of WDM channels with 50 GHz frequency separation is described by

$$193.1 + N \times 0.05 \text{ [THz]}, \qquad (2.28)$$

where N is an integer. The reference frequency is 193.1 THz, while the reference wavelength is 1552.52 nm. To avoid overlapping of signals, Δf should be smaller

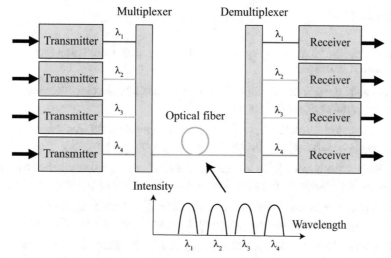

Fig. 2.14 Transmission system with WDM

than 50 GHz. Due to limited frequency selectability of the filters, the frequency allocation should have some margin between channels. For example, a WDM system with 50 GHz frequency separation can be used for optical signals with Δf smaller than 20 GHz. The ITU recommendation also defines WDM systems with frequency separations of 100, 25, and 12.5 GHz.

Table 2.1 Optical bands for WDM (in wavelength) [1]

T-band (thousand)	1000–1260 nm
O-band (original)	1260–1360 nm
E-band (extended)	1360–1460 nm
S-band (short wavelength)	1460–1530 nm
C-band (conventional)	1530–1565 nm
L-band (long wavelength)	1565–1625 nm
U-band (ultralong wavelength)	1625–1675 nm

Table 2.1 shows typical bands for optical fiber transmission. Optical amplifiers are required for long-haul transmission systems; therefore, the available bandwidth of WDM is limited by that of the amplifiers. C- and L-bands where low-cost and high-performance optical amplifiers are available are commonly used in commercial WDM transmission systems. S-, C-, and L-bands can provide over 400 WDM channels with 50-GHz optical frequency separation. The total optical power is 400 mW, when the optical power of each WDM channel is 1 mW. The optical power is limited by nonlinearity and damage threshold of SMFs.

Problems

2.1 Calculate the bit rate of an optical signal with a binary modulation format whose pulse period is 100 ps.

2.2 Calculate the bit rate of a 50 Gbaud signal with a multilevel modulation format, where the number of symbols is 64.

2.3 Calculate the bandwidth in wavelength for an optical signal whose bandwidth in frequency is 50 GHz, where the center wavelength is 1550 nm.

2.4 Calculate the bandwidth in wavelength for an optical signal whose bandwidth in frequency is 100 GHz, where the center wavelength is 1000 nm.

2.5 Calculate the number of optical channels in a WDM transmission system using the full span of T-band. The bandwidth of the optical channels is 100 GHz.

2.6 Calculate the total transmission capacity in a WDM transmission system using the full spans of S-band, C-band, and L-band. The bandwidth and baud rate of the optical channels are, respectively, 50 GHz and 40 Gbaud. The number of symbols in the modulator format is 16.

References

1. T. Kawanishi, *Wired and Wireless Seamless Access Systems for Public Infrastructure* (Artech House, Norwood, 2020)
2. A. Yariv, *Optical Electronics in Modern Communications*, 5th edn. (Oxford University Press, Oxford, 1996)
3. S.O. Kasap, *Optoelectronics and Photonics* (Prentice Hall, Englewood Cliffs, 2001)
4. T. Kawanishi, Integrated Mach-Zehnder interferometer-based modulators for advanced modulation formats, in *High Spectral Density Optical Communication Technologies, Optical and Fiber Communications Reports 6*, ed. by M. Nakazawa, K. Kikuchi, T.Miyazaki. (Wiley, Hoboken, 2010)
5. T. Kawanishi, High-speed optical communications using advanced modulation formats, in *Wiley Encyclopedia of Electrical and Electronics Engineering*, ed. by J.G. Webster (Wiley, Hoboken, 2016)
6. A. Sano, H. Masuda, Y. Kisaka, S. Aisawa, E. Yoshida, Y. Miyamoto, M. Koga, K. Hagimoto, T. Yamada, T. Furuta, H. Fukuyama, 14-Tb/s (140x111-Gb/s PDM/WDM) CSRZ-DQPSK transmission over 160km using 7-THz bandwidth extended L-band EDFAs. *European Conference on Optical Communications Proceedings (ECOC'06)* (2006)
7. A.H. Gnauck, G. Charlet, P. Tran, P.J. Winzer, C.R. Doerr, J.C. Centanni, E.C. Burrows, T. Kawanishi, T. Sakamoto, K. Higuma, 25.6-Tb/s WDM transmission of polarization-multiplexed RZ-DQPSK signals. J. Lightw. Technol. **26**(1), 79–84 (2008)

8. T. Kawanishi, Thz and photonic seamless communications. J. Lightw. Technol. **37**(7), 1671–1679 (2019)

9. T. Kawanishi, T. Sakamoto, M. Izutsu, High-speed control of lightwave amplitude, phase, and frequency by use of electrooptic effect. IEEE J. Sel. Topics Quant. Electron. **13**(1), 79–91 (2007)

10. M. Daikoku, I. Morita, H. Taga, H. Tanaka, T. Kawanishi, T. Sakamoto, T. Miyazaki, T. Fujita, 100-Gb/s DQPSK transmission experiment without OTDM for 100G Ethernet transport. J. Lightw. Technol. **25**(1), 139–145 (2007)

11. P.J. Winzer, G. Raybon, H. Song, A. Adamiecki, S. Corteselli, A.H. Gnauck, D.A. Fishman, C.R. Doerr, S. Chandrasekhar, L.L. Buhl, T.J. Xia, G. Wellbrock, W. Lee, B. Basch, T. Kawanishi, K. Higuma, Y. Painchaud, 100-gb/s DQPSK transmission: from laboratory experiments to field trials. J. Lightw. Technol. **26**(20), 3388–3402 (2008)

12. P.J. Winzer, A.H. Gnauck, 112-Gb/s polarization-multiplexed 16-QAM on a 25-GHz WDM grid, in *2008 34th European Conference on Optical Communication* (IEEE, Piscataway, 2008), pp. 1–2

13. T. Sakamoto, A. Chiba, T. Kawanishi, 50-Gb/s 16 QAM by a quad-parallel Mach-Zehnder modulator, in *33rd European Conference and Exhibition of Optical Communication - Post-Deadline Papers (published 2008)* (2007), pp. 1–2

14. T. Kawanishi, T. Sakamoto, M. Izutsu, K. Higuma, T. Fujita, S. Mori, S. Oikawa, J. Ichikawa, 40 Gbit/s versatile LiNbO$_3$ lightwave modulator, in *31st European Conference and Exhibition on Optical Communication (ECOC)* (2005)

15. T. Kawanishi, A. Kanno, H.S.C. Freire, Wired and wireless links to bridge networks: seamlessly connecting radio and optical technologies for 5G networks. IEEE Microw. Mag. **19**(3), 102–111 (2018)

16. C.H. Cox III, *Analog Optical Links* (Cambridge University Press, Cambridge, 2004)

17. V.J. Urick, Jr., J.D. McKinney, K.J. Williams, *Fundamentals of Microwave Photonics* (Wiley, Hoboken, 2015)

18. A. Kanno, K. Inagaki, I. Morohashi, T. Sakamoto, T. Kuri, I. Hosako, T. Kawanishi, Y. Yoshida, K. Kitayama, 40 Gb/s W-band (75-110 GHz) 16-qam radio-over-fiber signal generation and its wireless transmission. Opt. Express **19**(26), B56–B63 (2011)

19. P.S. Devgan, *Applications of Modern RF Photonics* (Artech House, Norwood, 2018)

20. H. Kiuchi, T. Kawanishi, M. Yamada, T. Sakamoto, M. Tsuchiya, J. Amagai, M. Izutsu, High extinction ratio Mach-Zehnder modulator applied to a highly stable optical signal generator. IEEE Trans. Microw. Theory Tech. **55**(9), 1964–1972 (2007)

21. T. Kawanishi, A. Kanno, P.T. Dat, T. Umezawa, N. Yamamoto, Photonic systems and devices for linear cell radar. Appl. Sci. **9**(3), 554 (2019)

22. S. Iezekiel, Measurement of microwave behavior in optical links. IEEE Microw. Mag. **9**(3), 100–120 (2008)

23. IEC 62802:2017, Measurement method of a half-wavelength voltage and a chirp parameter for Mach-Zehnder optical modulator in high-frequency radio on fibre (RoF) systems
24. IEC 62803:2016, Transmitting equipment for radiocommunication - Frequency response of optical-to-electric conversion device in high-frequency radio over fibre systems - Measurement method
25. K. Inagaki, T. Kawanishi, M. Izutsu, Optoelectronic frequency response measurement of photodiodes by using high-extinction ratio optical modulator. IEICE Electron. Express **9**(4), 220–226 (2012)
26. ITU-T G.694.1, Spectral grids for WDM applications: DWDM frequency grid, Recommendation

Chapter 3
Direct and External Modulation

Optical modulation techniques which modulate parameters of lightwaves are categorized into direct modulation and external modulation. Direct modulation, which changes some physical parameters of light sources, can generate intensity modulated signals with simple configurations. Laser diodes convert electric current into optical power. The output optical signal can be modulated by the change of the electric current. However, the linearity and high-speed response are limited by a finite threshold current. External modulation modulates continuous lightwaves by using external modulators. Various physical phenomena can be used for optical modulation. External modulation with electro-optic effect offers high-performance lightwave control, so that it is often utilized in high-speed or long-haul transmission, which requires precise and high-speed optical modulation.

3.1 Direct Modulation

In direct modulation, optical output power from a laser is controlled through electric current fed to lasers. The electric current is called injection current [1, 2]. For optical fiber transmission, laser diodes consisting of semiconductor layers are commonly used as light sources [3]. Laser diodes are also called as semiconductor lasers. Photons can be generated from holes and electrons injected into a p-n junction, where photon energy is dominated by the band gap of the semiconductor material. In actual laser diodes, holes and electrons are confined in a double heterostructure, for effective light emission. The emitted photons are also confined in a cavity for lightwaves.

Figure 3.1 shows a schematic of direct modulation, where the injection current consists of an average component and a small amplitude vibration component. The average component is called bias current. The vibration component is proportional to the electric modulating signal. As shown in Fig. 3.2, the optical output power can be controlled by the injection current (J). However, the optical power is not sensitive to the injection current, when the current is less than the threshold current (J_{th}).

© Springer Nature Switzerland AG 2022
T. Kawanishi, *Electro-optic Modulation for Photonic Networks*, Textbooks in
Telecommunication Engineering, https://doi.org/10.1007/978-3-030-86720-1_3

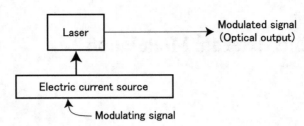

Fig. 3.1 Schematic of direct modulation

On the other hand, the change of the optical power is approximately proportional to the injection current, when J is much larger than J_{th}. The bias current should be larger than J_{th}, for effective modulation. When the bias point is close to J_{th}, the output profile would be distorted due to the nonlinear response around J_{th} shown in Fig. 3.2. To acheive linear response, the bias point should be set far from the threshold current J_{th}. However, when the average current is large, the power consumption and heating effect inside the laser diode would be an issue. In addition, it would be rather difficult to control large electric current quickly. Thus, the electric circuit for the current source would limit the frequency response of the direct modulation, if the bias current is much larger than the threshold current.

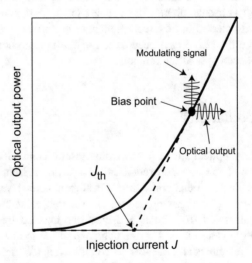

Fig. 3.2 Laser diode output power

Frequency response of a laser diode is also dominated by relaxation oscillation of lasers. Figure 3.3 shows a typical frequency response. The relaxation oscillation frequency, f_r, is given by

$$f_r = K_r\sqrt{\frac{J}{J_{th}} - 1},\qquad (3.1)$$

where K_r is a coefficient associated with overall bandwidth of the laser diode. The relaxation oscillation frequency is an increasing function of the ratio of the injection current J to the threshold current J_{th}. As mentioned above, it would be rather difficult to increase the injection current due to large power consumption. Thus, the reduction of J_{th} is very important to increase both power efficiency and bandwidth. J_{th}, which depends on the lasing process efficiency inside the laser diode, can be reduced by using a small volume device, or a low-loss cavity structure.

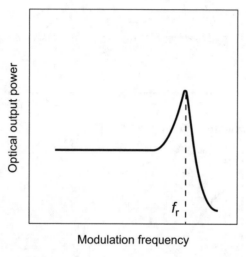

Fig. 3.3 Frequency response of direct modulation

In addition, direct modulation changes optical frequency as well as intensity. Thus, intensity modulated signals generated by direct modulation would have parasitic optical frequency or phase shift, as shown in Fig. 3.4. Such optical frequency or phase shift is called chirp [4, 5]. Frequency shift due to the chirp would be much larger than bandwidths of modulating signals, while external modulation can provide intensity modulation without chirp, as shown in Fig. 3.5.

Direct modulation, which can provide optical modulation with very simple configuration, is commonly used for short distance optical fiber links. However, it would be rather difficult to apply direct modulation to long-haul or large-capacity transmissions systems, because of response speed limitation and chirp effect. External modulation which will be described in the next section is commonly used for such transmission systems which require high-speed and precise control of lightwaves.

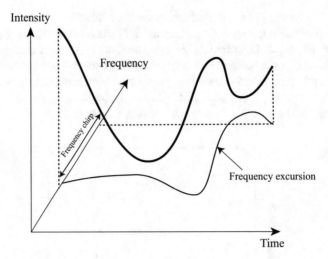

Fig. 3.4 Frequency excursion with chirp effect

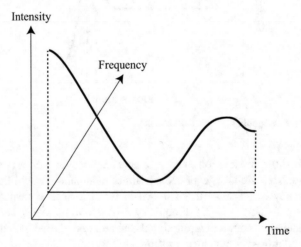

Fig. 3.5 Frequency excursion without chirp effect

3.2 External Modulation

In external modulation, external modulators vary some parameters of unmodulated continuous lightwaves [6–8]. The modulators can change amplitude, phase, or frequency quickly, according to modulating signals applied on electric input ports, as shown in Fig. 3.6. External modulation can provide precise and high-speed control of amplitude, phase, and frequency.

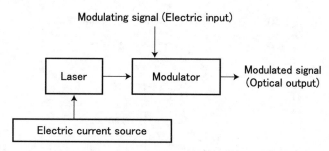

Fig. 3.6 Schematic of external modulation

3.2.1 Physical Effect for Modulation

Modulators utilize change of refractive indices or absorption coefficients in device materials whose physical parameters would depend on temperature, pressure, electric current, electric field, etc. Such effects induced by electric field can provide rapid response, so that electro-absorption (EA) and electro-optic effects (EO) are commonly used for high-speed operation [9, 10].

Magneto-optic (MO) and thermo-optic (TO) effects also can provide control of lightwaves through magnetic field and temperature. However, it would be rather difficult to realize rapid response. Thus, these effects are utilized for sensors and optical switches. Interaction between lightwaves and acoustic waves can be generated by acousto-optic (AO) effect. This effect which can respond acoustic waves whose frequency is up to a few hundred MHz is commonly used for measurement instruments.

3.2.2 Optical Modulation by EA Effect: EA Modulation

Semiconductor absorbs lightwaves whose energy is larger than bandgap. In particular semiconductor materials, the bandgap can be varied by intensity of electric field [9, 11]. This phenomenon is called Franz Keldysh effect. When the energy of the lightwave is close to the bandgap, absorptance would depend on that of the applied electric field, as shown in Fig. 3.7. Thus, the intensity of the output lightwave can be controlled by the voltage of the modulating signal. This is the basic principle of EA modulators. We can enhance such effect by using nano structures such as quantum wells [9, 11, 12]. This EA effect is called quantum confined Stark effect (QCSE).

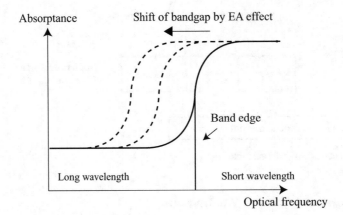

Fig. 3.7 Principle of EA modulation

3.2.3 Optical Modulation by EO Effect: EO Modulation

The EO effect would include various types of interaction between optical properties of materials and electric fields applied to materials [3, 6, 9, 10, 13]. However, the EO effect means refractive index change induced by electric field, in a narrow sense. The refractive index change would be proportional to the square or cubic of the electric field amplitude. For optical modulation, the Pockels effect, which is refractive index change proportional to the amplitude of the electric field, is commonly used to control optical phase, so that we will focus on this linear EO effect (Pockels effect), in this book.

Ferroelectric materials, such as lithium niobate LN : LiNbO$_3$, lithium tantalate LT : LiTaO$_3$, are commonly used as EO materials which can respond high-speed signals whose frequencies would be up to a few hundred GHz.

Semiconductor materials, whose bandgaps are close to the energy corresponding to 1.5 μm wavelength lightwaves which are commonly used for optical fiber networks, such as gallium arsenide (GaAs) and indium phosphide (InP), can offer high-speed EO modulation.

3.2.4 Intensity Modulation by Refractive Index Change

Refractive index change induces phase shift of lightwaves. For example, when the refractive index increases, time required to pass the modulator also increases. That causes phase lags. On the other hand, decrease of the refractive index provides phase leads. As described in the previous subsection, the EO effect can respond to very high-frequency signals. Thus, the modulator based on the EO effect can provide high-speed optical phase modulation. However, most of the optical communication systems use amplitude or intensity modulation formats, so that the phase lags or leads should be converted into change of intensity.

Optical interferometers, such as Michelson interferometer, Fabry-Perot interferometer, Mach-Zehnder interferometer, can convert small phase shift into intensity change. Mach-Zehnder (MZ) interferometer is commonly used for EO modulation, because we can easily fabricate MZ interferometers as photonic integrated circuits consisting of optical waveguides. The EO modulator with the MZ interferometer is called a Mach–Zehnder modulator (MZM), which provides wide wavelength operation. Details will be described in Sect. 4.3.

EA modulators also can provide high-speed optical intensity modulation, however, the output optical signals would have some chirp. On the other hand, the MZMs can provide pure amplitude modulation, in other word, zero-chirp modulation. Thus, the MZMs can be used for various types of modulation formats, which require precise control of phase and amplitude.

Problems

3.1 Consider a laser diode with an injection current J, where J is double the threshold current J_{th}. As shown in (3.1), the relaxation oscillation frequency, f_r is an increasing function of J. How much injection current is needed to increase f_r by a factor of 5?

3.2 Plot the relaxation oscillation frequency as a function of J_{th}.

3.3 Plot the relaxation oscillation frequency as a function of J.

References

1. R.G. Hunsperger, *Direct Modulation of Semiconductor Lasers* (Springer, New York, 2009)
2. Y. Matsui, D. Mahgerefteh, X. Zheng, C. Liao, Z.F. Fan, K. McCallion, P. Tayebati, Chirp-managed directly modulated laser (CML). IEEE Photon. Technol. Lett. **18**(2), 385–387 (2006)
3. T. Kawanishi, *Wired and Wireless Seamless Access Systems for Public Infrastructure* (Artech House, Norwood, 2020)
4. F. Koyama, K. Iga, Frequency chirping in external modulators. J. Lightw. Technol. **6**(1), 87–93 (1988)
5. T. Kawanishi, K. Kogo, S. Oikawa, M. Izutsu, Direct measurement of chirp parameters of high-speed Mach-Zehnder-type optical modulators. Opt. Commun. **195**(5–6), 399–404 (2001)
6. T. Kawanishi, T. Sakamoto, M. Izutsu, High-speed control of lightwave amplitude, phase, and frequency by use of electrooptic effect. IEEE J. Sel. Top. Quant. Electron. **13**(1), 79–91 (2007)

7. T. Kawanishi, Integrated Mach-Zehnder interferometer-based modulators for advanced modulation formats, in *High Spectral Density Opitcal Communication Technologies, Optical and Fiber Communications Reports 6*, ed. by M. Nakazawa, K. Kikuchi, T.Miyazaki (Wiley, Hoboken, 2010)
8. T. Kawanishi, High-speed optical communications using advanced modulation formats, in *Wiley Encyclopedia of Electrical and Electronics Engineering*, ed. by J.G. Webster (Wiley, Hoboken, 2016)
9. G.L. Li, P.K.L. Yu, Optical intensity modulators for digital and analog applications. J. Lightw. Technol. **21**(9), 2010–2030 (2003)
10. S.O. Kasap, *Optoelectronics and Photonics* (Prentice Hall, Englewood Cliffs, 2001)
11. K. Wakita, *Electroabsorption Effect* (Springer, Boston, 1998), pp. 79–111
12. Y. Ogiso, J. Ozaki, Y. Ueda, H. Wakita, M. Nagatani, H. Yamazaki, M. Nakamura, T. Kobayashi, S. Kanazawa, Y. Hashizume, H. Tanobe, N. Nunoya, M. Ida, Y. Miyamoto, M. Ishikawa, 80-GHz bandwidth and 1.5-V V_π InP-based IQ modulator. J. Lightw. Technol. **38**(2), 249–255 (2020)
13. A. Yariv, *Optical Electronics in Modern Communications*, 5th edn. (Oxford University Press, Oxford, 1996)

Chapter 4
Basics of Electro-Optic Modulators

This chapter describes basics of modulators based on EO effect, by using time domain mathematical expressions. In materials with electro-optic (EO) effect, the refractive index can be controlled by voltage applied to a modulator, so that optical phase can be controlled by the voltage. This is a basic principle of an optical phase modulator by EO effect. Various types of modulators can be constructed by using combinations of phase modulators. Intensity modulation can be offered by a Mach–Zehnder modulator (MZM) consisting of two phase modulators connected in parallel, where the optical phase difference between the two modulator is converted into optical amplitude change through a Mach–Zehnder interferometer (MZI). To achieve effective optical phase difference change, push-pull operation where the induced optical phases at the two phase modulators have the opposite signs to each other is commonly used. This chapter describes various device structures on LN substrates for high-speed operation, and configurations of electric circuits for push-pull operation. Vector modulation, which independently controls in phase and quadrature light components, can be provided by an integrated modulator consisting of phase modulators or MZMs. In actual modulators, optical phase fluctuation due to laser noise and mechanical vibration should be taken into account to ensure stable operation. Robustness against phase fluctuation is discussed in the last part of this chapter. In modulators consisting of two or more phase modulators, phase differences among the phase modulators connected in parallel should be fixed in integrated photonic circuit. Even in such integrated devices, slow phase shift would be induced by voltage applied to the modulator. This effect is called DC drift, which can be compensated by automatic bias control.

4.1 Mathematical Expression of Optical Modulation

This section offers mathematical expressions to describe various types of optical modulators. We focus on EO modulation with LN, which provides ideal EO effect. Operation of MZMs using lithium niobate (LN) can be described precisely by simple

© Springer Nature Switzerland AG 2022
T. Kawanishi, *Electro-optic Modulation for Photonic Networks*, Textbooks in
Telecommunication Engineering, https://doi.org/10.1007/978-3-030-86720-1_4

mathematical models. The models can be applied to semiconductor EO modulators, while semiconductor materials would have parasitic change of absorption coefficient.

As shown in Fig. 4.1, an optical modulator changes amplitude, phase, or frequency of an optical input, according to an electric modulating signal. In general, the optical input, optical output, and modulating signal can be expressed by scalar or vector variables. However, this book focuses on modulators with one optical input denoted by a scalar variable, Q, and with one optical output denoted by R. On the other hand, the modulating signal can be expressed by a scalar or vector variable.

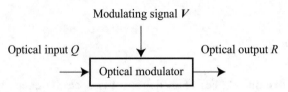

Fig. 4.1 Configuration of an optical modulator

A modulating signal with N components can be described by a N-dimensional vector, V, as follows:

$$V(t) = [V_1(t), V_2(t), \cdots, V_N(t).] \tag{4.1}$$

When the optical input is a monochromatic stable lightwave, the input can be expressed by,

$$Q = E_0 e^{i\omega_0 t} e^{i\Phi_0}, \tag{4.2}$$

where ω_0 and E_0 denote an angular frequency and amplitude of the optical input. The optical frequency and wavelength in a vacuum are given by $f_0 = \omega_0/2\pi$ and $\lambda_0 = c/f_0$, where c is the speed of light.

The optical input can be described by a phasor, $E_0 e^{i\Phi_0}$, where the term, $e^{i\omega_0 t}$, oscillating with the angular frequency, ω_0, defines the reference amplitude, phase, and frequency. Instead of the complex number expression shown in (4.2), real functions such as $\sin \omega_0 t$ and $\cos \omega_0 t$ can describe the input and output optical signals, where the optical modulation is a linear physical process for the optical input. The amplitude of the optical input does not affect the shape of the optical output spectrum, so that the output power is proportional to that of the optical input. On the other hand, the modulating signal should be expressed by a real function, because the modulation is a nonlinear process for the modulating signal. The spectrum would be changed by increasing the amplitude of the modulating signal. Details are described in Sect. 6.2.3.

The optical output (modulated signal) can be also expressed by a complex function, as follows:

$$R = KE(t) e^{i\omega_0 t} e^{i\Phi(t)}, \tag{4.3}$$

where $E(t)$ and $\Phi(t)$ denote the amplitude and phase of the modulated signal as a phasor. $E(t)$ and $\Phi(t)$ are functionals of the modulating signal. K corresponds to the intrinsic loss and phase delay during optical wave propagation in the modulator. $|K| \leq 1$ in common modulators where optical waveguides do not have any gain.

4.2 Optical Phase Modulation

Optical phase modulators can be configured by using EO effect which induces optical phase shift corresponding to applied electric fields. This section describes basics of optical phase modulation and practices of optical phase modulators. Optical phase modulators are the building blocks of various modulators. Knowledge of optical phase modulation is essential to understand these operating principles.

4.2.1 Electro-Optic Effect in Lithium Niobate

LN is ferroelectric and birefringent material, whose polarization direction is called polarization axis or optical axis. In general, crystal structures would be described by three axes (a, b, and c), where the c-axis is commonly chosen as the optical axis.

The refractive index depends on the direction of the electric field. A lightwave whose electric field is perpendicular to the c-axis is called an ordinary ray. On the other hand, an extraordinary ray has electric field parallel to the c-axis. The refractive index of the ordinary ray n_{0o} at wavelength of 1550 nm is 2.223, while that of the extraordinary ray n_{0e} is 2.143. The EO effect also depends on the direction of applied electric field and the electric field of the lightwave. In this section, the c-axis is assumed to be parallel to z-axis.

Refractive index change due to Pockels effect is commonly described by the inverse of the square of the refractive index, as follows:

$$\Delta\frac{1}{n_i^2} = r_{ij}E_j, \qquad (4.4)$$

where $n_i(i = 1, 2, 3)$ are refractive indices for x, y, and z direction electric components. $E_j(j = 1, 2, 3)$ are the x, y, and z direction components of the applied electric field. r_{ij} are called EO coefficients. r_{13}, r_{23}, and r_{33} denote EO effects on along x, y, and z axes, induced by the z direction component of the applied electric field. r_{11}, r_{21}, and r_{31} are due to the x direction component, while r_{12}, r_{22}, and r_{32} are due to the y direction component. For example, r_{33} shows the proportional constant of the EO effect on an optical signal whose electric field is along the z-direction, where the electric field of the modulating signal is also along z-direction (E_3). On the other hand, r_{31} denotes the EO effect on z-direction optical electric field with the modulating-signal electric field along the x-direction.

By using

$$\frac{d(\frac{1}{n_i^2})}{dn_i} = -\frac{2}{n_i^3},$$ (4.5)

the refractive index change due to Pockels effect (Δn_i) can be given by

$$\Delta n_i \simeq \Delta \frac{1}{n_i^2} \left[\frac{d(\frac{1}{n_i^2})}{dn_i} \right]^{-1}$$

$$= -\frac{n_i^3}{2} r_{ij} E_j.$$ (4.6)

The EO effect has a maximum when the directions of the applied electric field and that of the lightwave electric field are parallel to the c-axis. The EO effect in LN for extraordinary rays is given by an EO coefficient $r_{33} = 30.8 \times 10^{-12}$ m/V. The EO coefficient for the ordinary ray whose electric field is along the x-axis is $r_{13} = 8.6 \times 10^{-12}$ [m/V], where it is one-third of that for extraordinary ray. The required signal power of the modulator using r_{33} is approximately one-tenth of that of the modulator using r_{13}. Thus, LN modulators commonly utilize the EO effect of r_{33}, to reduce the required modulating-signal amplitude. To align the electric field parallel to the c-axis, the polarization of the optical signal should correspond to the extraordinary ray. For ordinary rays, the refractive index change induced by the EO effect is proportional to r_{13}, where $r_{23} = r_{13}$, due to the symmetry of the LN crystal structure.

Figure 4.2 shows a configuration of EO modulators utilizing the large EO coefficient for the extraordinary ray, where the EO effect is induced by the applied electric field proportional to the c-axis. Most of the EO modulators with LN are based on this configuration, to enhance modulation efficiency.

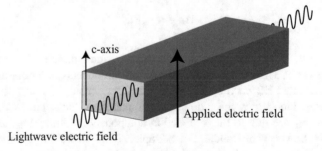

Fig. 4.2 EO effect for extraordinary ray

In this book, we focus on modulators based on the refractive index change induced by the EO effect associated with r_{33}, where the optical signals should be polarized in extraordinary rays. Frequency of an optical signal should be close to 200 THz, while that of a modulating signal is less than 100 GHz. The expected bandwidth for the modulating signal is wide enough for high-speed data transmission, however, the modulating frequency is much less than the optical frequency. Thus, the refractive index change can be considered quasi-static.

4.2.2 Principle of Optical Phase Modulation by Electro-Optic Effect

We assume that the optical input of the modulator is a monochromatic lightwave whose electric field Q is given by (4.2), where the amplitude, frequency, and phase are constant. The EO effect is a linear effect on the optical input; we can use phasors to describe lightwaves in the modulators. ω_0 and E_0 are angular frequency and amplitude of the optical input. The optical frequency can be given by $\lambda_0 = c/f_0$, where we used the equation connecting optical frequency and wavelength, $f_0 = \omega_0/2\pi$. On the other hand, we cannot use the phasor expressions for the modulating signal, because the EO modulation is not a linear phenomenon for the electric field applied to the modulator.

By using (4.6), we can obtain a mathematical expression of refractive index change due to the EO effect as follows,

$$\Delta n(t) = -n_0^3 r_{33} F(t)/2, \tag{4.7}$$

where the applied electric field and the electric field of the lightwave are assumed to be parallel to the c-axis. $F(t)$ is the amplitude of the applied electric field, which is proportional to the amplitude of modulating-signal voltage. The refractive index for the extraordinary ray whose electric field is parallel to the c-axis is expressed by

$$n = n_0 + \Delta n(t), \tag{4.8}$$

where the refractive index without any modulating signal is denoted by $n_0 = n_{0e}$.

Delay due to wave propagation in a modulator whose length of L can be written by

$$t_{\mathrm{DL}} = \frac{nL}{c}, \tag{4.9}$$

where nL is called an optical path length.

The output lightwave from the modulator (the modulated signal) would have phase delay relative to the input lightwave and can be expressed by

$$
\begin{aligned}
KE(t)\mathrm{e}^{i\omega_0 t}\mathrm{e}^{i\Phi(t)} &= K_L E_0 \exp\left[i\omega_0\left(t - \frac{nL}{c}\right) + i\Phi_0\right] \\
&= K_L E_0 \exp\left[i\omega_0\left(t - \frac{n_0 L}{c} - \frac{\Delta n(t)L}{c}\right) + i\Phi_0\right] \\
&= K_L \mathrm{e}^{i(\Phi_0 - \omega_0 n_0 L/c)} E_0 \mathrm{e}^{i\omega_0 t} \mathrm{e}^{-i\omega_0 \Delta n(t)L/c}
\end{aligned}
\tag{4.10}
$$

where optical loss in the modulator is denoted by $K_L{}^{-1}$ ($-20 \log K_L$ in decibels). K_L corresponds to optical transmittance of the modulator.

In this equation, the last factor

$$e^{-i\omega_0 \Delta n(t)L/c} = e^{i\omega_0 n_0^3 r_{33} F(t)L/2c} \tag{4.11}$$

depends on the induced electric field $F(t)$. The right side of this equation is derived from (4.7). The intensity of the output lightwave is constant, while the phase shift is proportional to the electric field $F(t)$.

By referring to (4.3), we get

$$E(t) = E_0 \tag{4.12}$$

$$K = K_L e^{i(\Phi_0 - \omega_0 n_0 L/c)} \tag{4.13}$$

$$\Phi(t) = -\omega_0 \Delta n(t)L/c$$
$$= -\frac{2\pi f_0 \Delta n(t)L}{c} = -\frac{2\pi \Delta n(t)L}{\lambda_0}. \tag{4.14}$$

This means that the EO effect affects only on the phase of the lightwave, where intrinsic loss and phase delay through wave propagation in the modulator can be expressed by K_L and $\omega_0 n_0 L/c$, respectively. Φ_0 is called an initial phase which expresses phase of lightwave at the input port of the modulator at $t = 0$. The initial phase depends on the definition of the time axis, so that the absolute value of the phase does not have physical significance. For simplicity, we use the time axis which satisfy

$$\Phi_0 = \omega_0 n_0 L/c. \tag{4.15}$$

By neglecting the optical propagation loss ($K_L = 1$), (4.10) can be rewritten by

$$R = E_0 e^{i\omega_0 t} e^{-i\omega_0 \Delta n(t)L/c}$$
$$= Q e^{-i\omega_0 \Delta n(t)L/c}. \tag{4.16}$$

This equation means that the EO effect changes the propagation speed of the lightwave in the modulator to modulate the phase of the output lightwave. Increase of refractive index $n(t)$ induces phase delay in the output lightwave, as shown in Fig. 4.3. In other words, a negative phase shift would be induced by a modulating signal with a positive voltage. This is the reason why the coefficient of the phase shift in this equation is negative. When $\Delta n(t)L$, which denotes optical path length shift due to the EO effect, is equal to a half wavelength, we can obtain a polarity inverted waveform. The refractive index change Δn ranges from 10^{-5} to 10^{-4}, and device length of a common LN modulator is a few centimeters, where the wavelength in LN is shorter than 1μm. Thus, $\Delta n(t)L$ can be set to the half wavelength of the optical signal inside the LN modulator.

Various types of optical modulation schemes can be constructed by the use of optical phase modulation described by (4.16) and (4.14). For example, we can configure an intensity modulator by using a pair of phase modulators connected in parallel, as shown in Sect. 4.3.1. Intensity or amplitude of the optical output

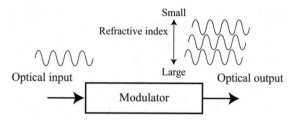

Fig. 4.3 Principle of optical phase modulation by EO effect

depends on relative optical phase difference between optical signals from the two optical phase modulators. Phase difference stability is very important in such optical modulators consisting of parallel optical phase modulators. A lightwave from a laser source has intrinsic phase fluctuation, so that it would be rather difficult to define an absolute phase of an optical signal. Figure 4.4 shows a typical optical phase noise power spectral density (PSD), where major part of the noise power is in the low frequency range. A laser output spectrum has a finite linewidth due to such optical phase fluctuation, as shown in Fig. 4.5. In a common semiconductor laser diode, the linewidth is about a few MHz. By using an external cavity, the optical phase fluctuation can be suppressed, where the linewidth can be less than a few hundred kHz.

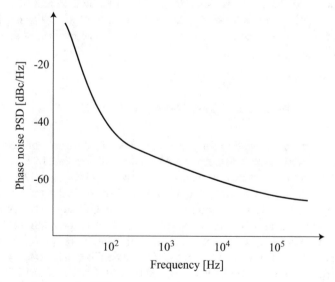

Fig. 4.4 A typical phase noise PSD

Mechanical vibration and temperature change over optical fibers also cause optical phase fluctuation. Optical phase delay of a lightwave propagating along an optical waveguide in an LN modulator would shift slowly due to mobile ions. This is called

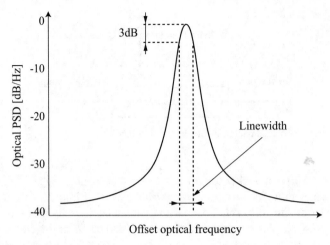

Fig. 4.5 Laser linewidth

DC drift, which causes fluctuation of optical output amplitude and phase. These effects should be compensated to offer stable optical communication transmission. Section 4.5 describes details on optical phase stability issues. In general, DC stands for direct current. However, in principle, no current is required to control optical phase through the EO effect. The electric field induced by the DC voltage changes the refractive index. In this book, DC refers constant voltage or bias condition. A DC voltage is applied to the electrode to control the bias condition as described in Sect. 4.3.3.

4.2.3 Structure of Optical Phase Modulator

Figure 4.6 shows a schematic of a phase modulator (PM) using an LN substrate with an optical waveguide. The electric field of a modulating signal fed to electrodes along with the waveguide induces change of refractive index by the EO effect. As described in Sect. 4.2.2, we can control the phase of the optical signal at the output port by the voltage of the modulating signal. Titanium diffusion is commonly used to form waveguides on LN substrates, where titanium is thermally diffused into the substrates to enhance refractive index under titanium strips [1–6]. As shown in Fig. 4.7, the waveguide is placed where the electric field is concentrated, to enhance the EO effect. The mode size of the waveguide is similar to that of SMFs, so that we can obtain low-loss connections with fibers easily.

The electric field $F(t)$ which is proportional to the voltage of the modulating signal $V_1(t)$ can be given by $F(t) = UV_1(t)$, where U is the proportionality factor. By using (4.7), the refractive index can be rewritten by

$$\Delta n(t) = WV_1(t), \tag{4.17}$$

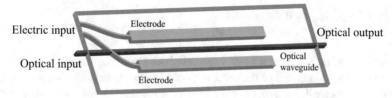

Fig. 4.6 Overview of an optical phase modulator

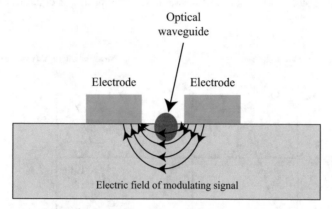

Fig. 4.7 Cross section of an optical phase modulator

where W is defined by

$$W \equiv -\frac{n_0^3 r_{33} U}{2}, \qquad (4.18)$$

and is called the modulating efficiency in this book. W denotes optical phase modulation efficiency per unit length of the electrode.

By using Γ defined by

$$\Gamma \equiv -\frac{2\pi L W}{\lambda_0}, \qquad (4.19)$$

the optical output of the modulator is given by

$$R = K_L E_0 e^{i\omega_0 t + iv_1(t)} \qquad (4.20)$$

$$v_1(t) \equiv \Gamma V_1(t). \qquad (4.21)$$

$v_1(t)$, which is optical phase shift induced by the modulating signal $V_1(t)$, is also called a modulating signal henceforth $v_1(t)$.

The coefficient, Γ, which expresses interaction between the modulating-signal electric field and the lightwave electric field, depends on the frequency of the modulating signal. Thus, the frequency response should be taken into account to consider modulation with high-frequency signals. As described in (4.1), the modulating signal should be expressed by a vector variable, to consider an integrated modulator consisting of two or more phase modulators.

The voltage $V_{\pi\mathrm{PM}}$ defined by

$$V_{\pi\mathrm{PM}} \equiv \frac{\pi}{\Gamma} \tag{4.22}$$

denotes voltage required for π (180°) optical phase shift. $V_{\pi\mathrm{PM}}$ is called a half-wave voltage of PM. This is one of the most important index to describe performance of modulators. Figure 4.8 shows optical phase shift induced by the modulating signal fed to the PM. When the zero-to-peak signal amplitude ($V_{0\mathrm{p}}$) is equal to $V_{\pi\mathrm{PM}}$, the optical phase shift, $\Phi(t)$, oscillates between $\pm\pi$. The period is equal to that of the modulating signal

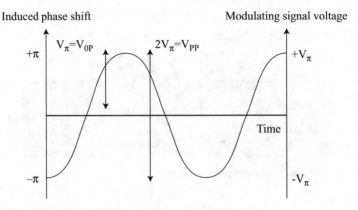

Fig. 4.8 Optical phase shift induced by a half-wave voltage signal

If the electric circuits are composed of transmission lines whose characteristic impedance equals 50 Ω, the required signal power is given by

$$P_\pi = \frac{V_{\pi\mathrm{PM}}^{\,2}}{100} \quad [W]. \tag{4.23}$$

Low $V_{\pi\mathrm{PM}}$ modulator can provide effective optical phase modulation with small amplitude electric inputs. When the overlaps of the electric fields of the modulating signal and lightwave is large, W would be large to reduce the half-wave voltage. To reduce the half-wave voltage, the optical waveguide can be placed in the area where the electric field is concentrated.

Figure 4.9 shows a cross section of a modulator with a coplanar waveguide (CPW) as a modulating-signal electrode. A CPW consisting of a signal electrode placed between two ground electrodes can offer low-loss signal transmission. CPWs are commonly used for commercial LN modulators, where the signal frequency can be up to 100 GHz. We can configure effective connection between a CPW and a coaxial cable, because both the signal and ground electrodes are on the top surface of the EO material.

Figure 4.10 shows a schematic of a modulator with a microstrip line (MSL) as an example. An MSL has a ground plane on a bottom surface. We can configure branch structures easily, because MSL can offer less complicated configurations than in CPW. Thus, MSL are often used signal transmission inside integrated devices. In both cases (CPW and MSL), the electric field of the modulating signal would be concentrated at the area close to the edge of the signal electrode, where the direction of the electric field is approximately perpendicular to the top surface. Thus, to achieve effective EO effect, the overlaps of the electric fields of the modulating signal and lightwave can be increased by placing the optical waveguide close to the edge of the signal electrode, where the input lightwave should be in TM-polarization where the electric filed is perpendicular to the top surface. The c-axis is also perpendicular to the top surface, to align with the electric field induced by the modulating signal. Such a substrate is called a z-cut substrate.

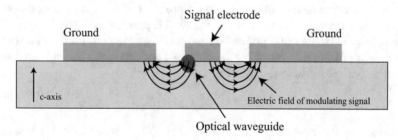

Fig. 4.9 Cross section of an optical phase modulator using a coplanar waveguide

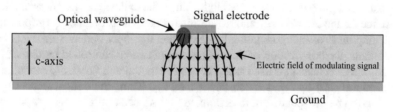

Fig. 4.10 Cross section of an optical phase modulator using a microstrip line

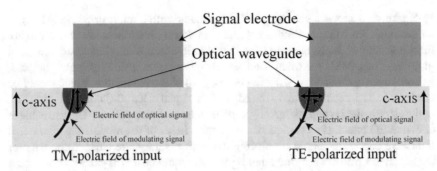

Fig. 4.11 Electric fields of modulating signal and optical signal in a z-cut optical phase modulator

Figure 4.11 shows an enlarged view close to the optical waveguide. The left figure depicts the electric field of the modulating signal that of the TM-polarized optical signal, while the right figure shows those for TE-polarization optical input. The electric fields of the optical signal and the modulating signal, and the direction of the c-axis are parallel to each other. For TE-polarized optical signals in z-cut substrates, the refractive index change described by

$$\Delta n(t) = -n_{0o}^3 r_{13} F(t)/2 \tag{4.24}$$

is induced by the modulating signal $F(t)$. The EO coefficient for TE-polarized component is given by r_{13}. On the other hand, that for TM-polarized component is r_{33}, which is about three times as large as r_{13}.

As shown in (4.7), EO coefficients of refractive index change are different for TE-polarized light and TM-polarized light. Thus, the optical path length would have difference for the ordinary ray and extraordinary ray. The difference is proportional to the amplitude of the modulating signal. When the input optical signal has both components, the polarization of the optical output can be controlled by the applied voltage. This mechanism is used as a polarization modulator. For phase modulation, optical phase shift by r_{33} is much larger than r_{13}. Thus, the optical input should be aligned to be TM-polarized, to maximize the induced optical phase. A polarizer passing TM-polarized light would be placed at the optical output port to suppress undesired components.

In this book, we will henceforth consider the EO modulation induced by the electric field along the c-axis. Figure 4.12 is a schematic of modulating signal electric field induced by the modulating signal, in a z-cut LN substrate. The electric field amplitude along the c-axis is also shown in the lower plot, where the electric field intensity has peaks under the edges of the ground and signal electrode (A, C, D, and F).

As described in Sect. 4.3.1, push-pull operation of phase modulators connected in parallel should be required to configure an MZM for amplitude or intensity modulation. For example, we can achieve effective push-pull operation by placing two optical waveguides at A and C in Fig. 4.12, where the signs of the electric field amplitudes are opposite to each other. However, the absolute values of the electric field amplitudes have some difference, which cause undesired phase shift in amplitude modulation. On the other hand, x-cut substrates whose c-axis is perpendicular to the normal vector of the top surface and to the optical wave propagation direction, in other words, is horizontal in Fig. 4.12 are often used, to achieve balanced optical phase modulation by two or more parallel phase modulators. As shown in Fig. 4.13, the electric field intensity in an x-cut LN substrate has peaks between the ground and signal electrode. The absolute values of electric fields at B and E are equal and the signs are opposite. More details on modulators using x-cut substrates will be described in Sect. 4.3.6. For a discrete phase modulator, an optical waveguide should be placed under the edge of the signal electrode in the z-cut substrate (point C or D in Fig. 4.12), where the electric field intensity is maximum.

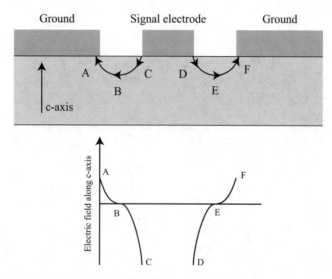

Fig. 4.12 Modulating-signal electric field inside z-cut LN substrate

4.2.4 Traveling-Wave Electrode for High-Speed Operation

The half-wave voltage $V_{\pi_{PM}}$ can be reduced by increasing the length of the modulator L, which the EO modulation efficiency coefficient, Γ, is proportional to, as described in (4.19). The induced optical phase can be enhanced by using a long electrode without increasing the voltage of the electric input, when loss in the electrodes is neglected. However, due to the limitation of the high-speed response of a series circuit consisting of a capacitor and a resistor, the voltage effectively applied to the electrodes would be small when the frequency of the modulating signal is high. A pair of conductors form a capacitor, where the driving circuit and the electrodes have finite resistance. If the length of the electrode is longer than the wavelength of the electric signal applied on the electrode, the shape of the electrode should be designed by using distributed constant element theory. The EO effect would be accumulated through lightwave propagation in the waveguide.

In a typical LN modulator, to achieve a half-wave voltage less than a few volts, the electrode length L should be longer than a few centimeters, because the EO coefficient r_{33} is not so large. On the other hand, the cavity length of laser diodes and the width of photodetectors are less than a few hundreds of microns. Electrodes in such small devices can be described by a lumped-element model, shown in Fig. 4.14. R_a is total resistance in the circuit, while C_a denotes the capacitance between the signal and ground electrodes. The voltage effectively induced between the electrodes can be described by

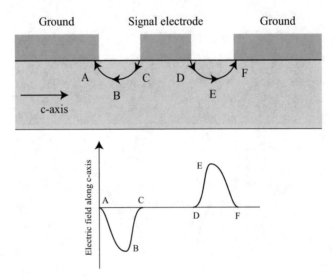

Fig. 4.13 Modulating-signal electric field inside X-cut LN substrate

$$V_{\text{eff}} = V_{\text{in}} \frac{1}{\sqrt{1 + (2\pi f R_a C_a)^2}}, \qquad (4.25)$$

where V_{in} and f are, respectively, voltage and frequency of an input sinusoidal signal. V_{eff} is a decreasing function of the frequency f. When

$$f = \frac{1}{2\pi R_a C_a} \equiv f_c, \qquad (4.26)$$

the signal power effectively applied to the electrode is half the input signal power, where

$$V_{\text{eff}} = V_{\text{in}}/\sqrt{2}. \qquad (4.27)$$

f_c, which is called cutoff frequency, describes electrode bandwidth. To increase f_c, the resistance R_a and capacitance C_a should be minimized.

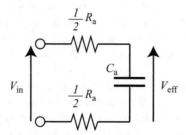

Fig. 4.14 Lumped-element circuit model

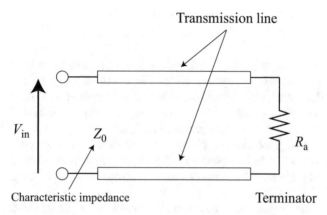

Fig. 4.15 Transmission line model

However, there is a trade-off between the bandwidth (f_c) and modulation efficiency (Γ), because C_a is proportional to L. As described above, electrodes longer than a few centimeters are needed to offer effective EO modulation, where the capacitance between the signal and ground electrodes becomes large. The cutoff frequency would be less than a few GHz. To overcome this difficulty, we should use transmission lines such as CPW and MSL shown in Figs. 4.9 and 4.10, as electrodes. Figure 4.15 shows a circuit model for a transmission line. To suppress undesired reflection, the circuit should be terminated by a resister whose resistance R_a is equal to the characteristic impedance of the transmission line Z_0. The voltage amplitude does not depend on the position on the transmission line, if the loss of the transmission line can be neglected. However, actual transmission lines should have some propagation loss due to conductor loss in electrodes and dielectric loss in substrate. Thus, the modulating signal propagates along the transmission line with exponential decay. Such decay limits the modulation efficiency in high-frequency region. In actual LN modulators using transmission lines, the impedance at the electric input port can be large enough even in high-frequency regions up to 100 GHz, while the impedance decreases rapidly in the capacitance-resistance series circuit.

Figure 4.16 shows a schematic of an optical modulator with a transmission line (CPW), where the modulating signal on the electrode and the optical signal in the optical waveguide propagate along the same direction. An electrode forms transmission line placed over an optical waveguide is called a traveling-wave electrode, which is commonly used to enhance the modulation bandwidth and to reduce the half-wave voltage [7, 8]. In the traveling-wave electrode, the speed of the modulating signal should be close to that of the optical signal, to accumulate induced optical phase effectively. Here, we consider the cutoff frequency caused by propagation speed difference between the optical and electric signals. As shown in Fig. 4.17, electric and optical signals propagate in a phase modulator in different velocities. The propagating velocities of the electric signal on the electrode, v_m, and that of the optical signal in the waveguide, v_0, are described by

$$v_m = c/n_m \tag{4.28}$$

$$v_0 = c/n_0, \tag{4.29}$$

where n_m and n_0 are, respectively, effective refractive indices for electric and optical signals. In general, the propagation velocity of the modulating signal is not equal to that of the optical signal, because effective refractive indices n_m and n_0 have difference. The electric signal propagates along a transmission line consisting of electrodes (conductor) and substrate (dielectric), while the optical signal is confined in an optical waveguide (dielectric). The refractive index depends on frequency of the propagating wave. In LN substrates, that for an optical wave whose frequency is close to 200 THz is much less than that for an electric signal whose frequency is less than 100 GHz.

Difference in propagation time at the modulator whose length is L can be expressed by

$$\tau = \left| \frac{L}{v_m} - \frac{L}{v_0} \right| = \frac{|n_m - n_0|L}{c}. \tag{4.30}$$

When the half-period of the signal is less than the difference, τ, the phase misalignment at the optical output would be larger than π. Thus, the signal whose period is close to or less than 2τ would have significant phase excursion during the modulation process. Here, we define the cutoff frequency due to this effect, as follows:

$$f_T = \left(\frac{2}{\tau} \right)^{-1} = \frac{c}{2|n_m - n_0|L}. \tag{4.31}$$

The bandwidth can be expanded by reducing the propagation speed difference, $|n_m - n_0|$, where the half-wave voltage can be reduced by using large L, [4, 5, 7, 8]. The response of the EO effect in LN is fast enough for modulating signals whose frequency is less than a few hundred GHz, so that the frequency dependence of the EO effect is negligible. A more rigorous analysis on the impact of the propagation velocity difference was given in section 6.3.1.2 of Ref. [9], where the cutoff frequency can be expressed by

$$f_{\mathrm{T}} = \frac{1.9c}{\pi|n_m - n_0|L} \simeq 0.6\frac{c}{|n_m - n_0|L}. \tag{4.32}$$

This is consistent with the approximate model shown in (4.31).

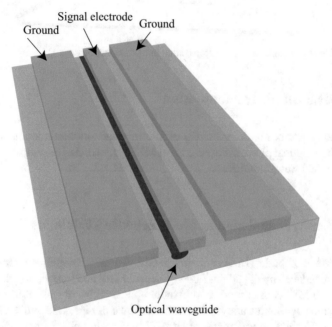

Fig. 4.16 Optical phase modulator using a travelling-wave electrode

In a traveling-wave electrode with the velocity matching condition: $n_{\mathrm{m}} = n_0$, the modulation efficiency is independent from the modulating-signal frequency, if the propagation loss on the transmission line is neglected. However, as described above, the bandwidth is also limited by the propagation loss of the electric signal on the electrode as well. The electric signal loss is an increasing function of the frequency, while the optical loss in the optical waveguide is independent from the modulation signal frequency. When the electric signal frequency is less than 100 GHz, the conductor loss is dominant in the propagation loss. However, the dielectric loss for signals whose frequency is higher than 100 GHz would have significant impact on the modulating efficiency.

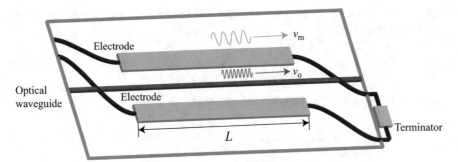

Fig. 4.17 Electric and optical signals propagating in a phase modulator

4.3 Optical Intensity Modulation

This section describes basic principles of intensity and amplitude modulation by an MZM, where optical phase difference in an MZI is converted into amplitude change through optical wave interference.

4.3.1 Intensity Modulation by Mach-Zehnder Modulators

MZIs, or Mach–Zehnder (MZ) structures in other words, are widely used for intensity modulation induced by EO effect. Ludwig Zehnder [10] and Ludwig Mach [11] were proposed the MZI consisting of two parallel optical paths, or optical arms in other words, formed by mirrors and half mirrors as shown in Figs. 4.18 and 4.19.

Figure 4.20 shows a schematic of an MZI, where an optical input fed from the input port A is split evenly by a half-mirror, reoriented by a mirror, and then merged again by a half-mirror. The operation principle is that the optical input is divided into two parts and then combined. The intensity of the optical output would be changed by optical interference which occurs at the half-mirror which combines the two parts. The intensity depends on the optical phase difference between the two parts on the two optical paths. This interferometer was used to detect the difference in optical phase change when passing through different samples.

By putting two samples on the two paths, the optical phase difference can be obtained from the optical power of the optical output C, which is the optical output at the port C, as shown in Fig. 4.20. When the optical phase difference is 0, the power of the optical output C would be maximum. On the other hand, when the optical phase difference is π (180°), the intensity would be minimum. As interferometers have no mechanism to absorb optical waves, the sum of the optical powers at output ports C and D is constant. When the optical output C goes to minimum, the optical output D which propagates along the dashed line through the port D goes to maximum. As shown in Fig. 4.20, the phase difference can be converted into the change of the output optical power, by using the optical output C as an output of the interferometer.

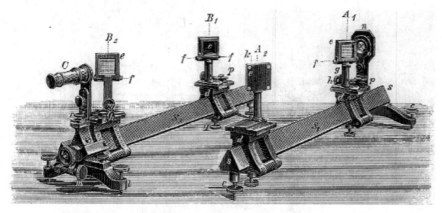

Fig. 4.18 Optical interferometer proposed by Zehnder [10]

Fig. 4.19 Optical interferometer proposed by Mach [11]

In addition, we may input an optical signal from the port B. Here, we consider a phase difference which maximize the optical output C induced from an optical input from the port A, or to minimize the optical output D in other words. An optical input from the port B also generates optical outputs from the ports C and D. The optical outputs C and D would be, respectively, minimum and maximum, with the phase difference which maximize the output C induced from the optical input A.

The behavior of the interferometer can be explained by the process in which independent horizontally and vertically propagating light waves are combined by a half-mirror on the input side, mixed in the interferometer, and also converted to horizontally and vertically propagating light waves independently on the output side.

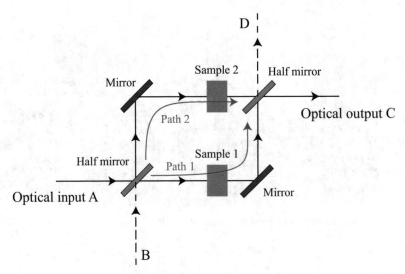

Fig. 4.20 Principle of Mach-Zehnder interferometer

Since nonlinear phenomena are not included in the interferometer in principle, and the input light consisting of two components is transformed into the output light consisting of two components, the function of the interferometer can be described by a matrix with two rows and two columns.

For optical intensity modulation, the optical phase difference is controlled by voltage through the EO effect. The phase change obtained by the EO effect is converted to an intensity change or amplitude modulation by optical interference, and is called a Mach–Zehnder modulator (MZM). As shown in Fig. 4.21, the MZM consists of two optical phase modulators which are integrated in parallel through two Y-junctions. The light waves from the two phase modulators are combined into an optical output. The two light waves interfere with each other at the Y-junction for the optical output, where the optical phase difference induced by the two phase modulator is converted into optical intensity change. Thus, the intensity of the optical output can be controlled by the voltages applied to the phase modulators. Figure 4.22 shows the principle of operation of the MZM. When the lights from the two phase modulators are in phase, or the optical phase difference is zero, in other words, the output optical power becomes a maximum, due to constructive interference at the Y-junction for the output. This is called the on-state. On the other hand, when the phase difference between the two optical signals from the two phase modulators is π (180°), the output optical power becomes a minimum, due to destructive interference. This is called the off-state. The energy of the light waves converted to radiation mode waves spread outside the waveguide at the combining section on the optical output side, and the optical intensity becomes zero at the optical output connected to the optical fiber.

In the MZI shown in Fig. 4.20, mirrors are used to change the propagation direction of light waves, allowing the branched light waves to merge back together.

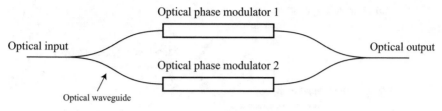

Fig. 4.21 MZM consisting of a pair of optical phase modulators

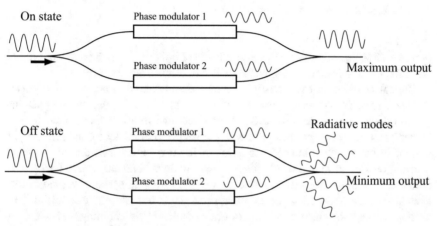

Fig. 4.22 Principle of intensity modulation with an MZM

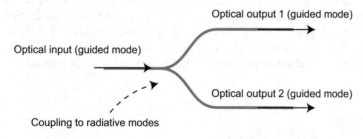

Fig. 4.23 Light wave split by a Y-junction

In the MZM, however, the optical waveguides play this role. In addition, optical branching and combining are realized by Y-junctions formed by optical waveguides. While the half-mirror was a two-input, two-output device, a Y-junction is one-input, two-output or two-input, one-output devices. However, as shown in Figs. 4.23 and 4.24, if we take into account the fact that the Y-junction causes coupling to radiative modes on the side with fewer connections (the output side in the case of mixing and the input side in the case of splitting), the Y-junction can be expressed as a 2-input/2-output device. In other words, an input consisting of one waveguide mode and one radiation mode is connected to an output consisting of two waveguide modes, or vice versa, which is mathematically equivalent to a half-mirror [12].

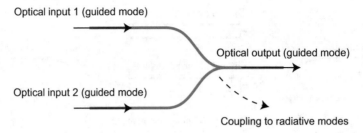

Fig. 4.24 Light wave mixing by a Y-junction

Figure 4.25 schematically shows electric fields of light waves in the two waveguides when the two light waves are in phase (phase difference equals zero).

Optical waveguides that satisfy the single-mode condition are commonly used to form MZIs. As shown in Fig. 4.25, at the point where the two waveguides begin to overlap, the waveguide width is wide enough to allow the propagation of higher-order mode components over a short distance (multi-mode condition). In the case of in phase, when the electric fields of the two light waves are combined, a distribution similar to that of the fundamental mode is obtained, where the electric field intensity is the maximum at the center of the waveguide. The waveguide width becomes narrower toward the output side, satisfying the single-mode condition. The overlap between the electric field distribution obtained by combining waves and the fundamental mode of the waveguide is large, and the waveguide mode is efficiently excited. In the case of reverse phase (phase difference π), as shown in Fig. 4.26, the electric fields of the two light waves are combined to form a distribution corresponding to a higher-order mode with the light intensity minimized at the center of the waveguide. On the way to the output side of the waveguide, it is converted to radiation mode, and the output from the waveguide becomes the minimum.

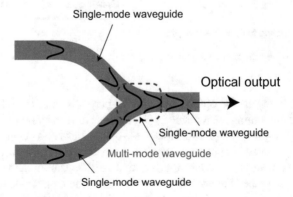

Fig. 4.25 Light wave interference at a Y-junction (on-state)

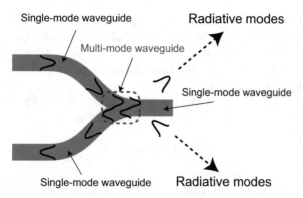

Fig. 4.26 Light wave interference at a Y-junction (off-state)

In the MZM, the optical phases on the two arms can be controlled by the modulation signal $\mathbf{V}(t) = [V_1(t), V_2(t)]$ applied to the optical phase modulators 1 and 2, and the output optical intensity can be changed through the light wave interference at the Y-junction for the output. The optical phase change induced by each phase modulator $v_i(t)$ is given by

$$v_i(t) = \Gamma_i V_i(t). \tag{4.33}$$

The optical output can be expressed by

$$R = E_0 e^{i\omega_0 t} \left[K_1 e^{iv_1(t)} + K_2 e^{iv_2(t)} \right]. \tag{4.34}$$

K_i describes optical transmittance of the optical phase modulator $i, (i = 1$ or 2). If optical loss in the phase modulator is neglected, K_i is equal to unity.

K_i also includes the effect of the amplitude change in the light wave split and mixing at Y-junctions. Even if we ignore optical loss due to material absorption in the Y-junctions, imperfections in the device structure, etc., the light wave split and mixing induce intrinsic amplitude change [12]. As shown in Fig. 4.27, a balanced Y-junction splits an optical input whose amplitude is E_0 into two outputs with amplitude of $E_0/\sqrt{2}$, where the input optical power is equally divided into two outputs.

Here, we consider light wave mixing by a balanced Y-junction. When light of amplitude $E_0/\sqrt{2}$ is input from one of the waveguides, the amplitude is further multiplied by $1/\sqrt{2}$ to obtain an optical output of $E_0/2$, as shown in Fig. 4.28. In this case, half of the light energy is lost, which is due to the fact that it is converted to the radiation mode in the off-state described above and dissipated outside the waveguide. In the Y-junction for mixing, when the optical input is from only one side, the energy is always distributed to the optical output and radiation modes, so the optical signal cannot be transmitted to the output side without loss. Only when optical signals of equal amplitude are input from both ports in the same phase, the interference effect cancels out the radiation and all the optical energy can be concentrated in the output port.

To summarize the above discussion, the output amplitude of the MZ structure is half the input amplitude, because the amplitude transmittance of the Y-junction for splitting is $1/\sqrt{2}$ and that for mixing is also $1/\sqrt{2}$. $K_i = 1/2$ for an ideal well-balanced MZ structure when loss due to material absorption or imperfections of the device structure is negligibly small.

An ideal intensity modulation can be achieved with an MZM, when the magnitude of the phase change in the two phase modulators is the same, but the signs are opposite, that is,

$$v_2(t) = -v_1(t). \tag{4.35}$$

This is called push-pull operation.

When $v_1(t) = -v_2(t) = g(t)$, $2g(t)$ shows the optical phase difference between the two light waves in the two arms. If the optical loss in the two phase modulators is also well-balanced, or $K_1 = K_2 = K/2$, in other words, the optical output can be given by

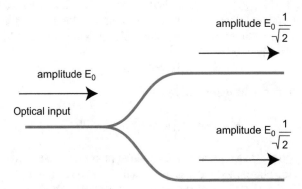

Fig. 4.27 Amplitudes of splitted light waves in a Y-junction

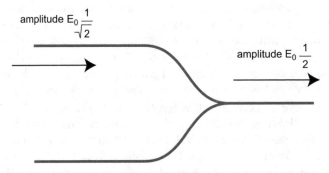

Fig. 4.28 Amplitudes of mixed light waves in a Y-junction

$$R = \frac{KE_0}{2}\left[e^{ig(t)} + e^{-ig(t)}\right]e^{i\omega_0 t}$$
$$= KE_0 e^{i\omega_0 t}\cos\left[g(t)\right]$$
$$= K\cos\left[g(t)\right]Qe^{-i\Phi_0}, \tag{4.36}$$

where we used Euler's formula, $e^{i\theta} = \cos\theta + i\sin\theta$.

In general, $|K|$ is less than unity, however, $K = 1$ if optical loss due to scattering, material absorption, etc. can be neglected. $e^{-i\Phi_0}$ shows a constant phase change due to propagation delay in the MZM. Φ_0, which is not time variant, does not directly related to optical modulation function.

By referring to (4.3), we get

$$E(t) = E_0 \cos\left[g(t)\right] \tag{4.37}$$
$$\Phi(t) = 0, \tag{4.38}$$

which means that an MZM with the push-pull operation realizes the function of amplitude modulation by multiplying the amplitude of the input light by $\cos\left[g(t)\right]$. The intensity of the output light relative to the input light is given by the following equation:

$$|R/Q|^2 = |K\cos\left[g(t)\right]|^2$$
$$= K^2\frac{1 + \cos\left[2g(t)\right]}{2}. \tag{4.39}$$

When the optical phase difference is zero, that is,

$$2g(t) = 2m\pi \quad m = \cdots, -1, 0, +1, \cdots, \tag{4.40}$$

the modulator is in the on-state. When

$$2g(t) = (2m + 1)\pi, \tag{4.41}$$

the modulator is in the off-state. Half-wave voltage of an MZM, $V_{\pi\mathrm{MZM}}$, can be defined by the voltage required to change the modulator from the on-state to the off-state.

As shown in Fig. 4.29, the half-wave voltage $V_{\pi\mathrm{MZM}}$ of the MZM corresponds to half of the half-wave voltage $V_{\pi\mathrm{PM}}$ of each phase modulator integrated in the MZM. In ideal push-pull operation, the $\pi/2$ phase changes at each phase modulator are added together. As a result, the phase difference between the light waves of the two waveguides changes by π and the switching of the MZM between the on-state and off-state is realized.

In commercial modulators, the definition of half-wave voltage is the same as above, but note that both $V_{\pi\mathrm{PM}}$ and $V_{\pi\mathrm{MZM}}$ are often described simply as half-wave voltage V_π. In a phase modulator, the main purpose is to control the output optical phase, and the half-wave voltage $V_{\pi\mathrm{PM}}$ is defined as the voltage required to shift the optical output phase by π, that is, by half a wavelength. On the other hand, in the

MZM, the main purpose is to change the intensity of the output light, and since the output light intensity depends on the optical phase difference between the two phase modulators, the voltage corresponding to the π change in this phase difference is defined as the half-wave voltage $V_{\pi \text{MZM}}$.

4.3.2 Amplitude Modulation and Parasitic Phase Modulation

As described in (4.40), when an MZM is in an on-state of $g(t) = m\pi$, $|\cos[g(t)]| = 1$. Thus, the optical intensity of the output $|R|$ becomes a maximum and does not depend on m. However, the polarity of R depends on m. When m is an odd number, the polarity is negative. On the other hand, when m is an even number, the polarity is positive. By continuously varying $g(t)$ from $2m\pi$ to $2(m + 1)\pi$ as shown in (4.36), the amplitude of the output light can be controlled arbitrarily. For example, when $g(t) = 0$, $\cos[g(t)] = +1$, but when $g(t) = \pi$, $\cos[g(t)] = -1$.

In other words, the MZM reverses the sign of the light wave according to the modulation signal. Although the MZM is widely used as an intensity modulator, it is essentially an amplitude modulator and can be applied to binary-phase-shift-keying (BPSK). This is due to the fact that BPSK is essentially a binary amplitude-shift-keying (ASK) with ± 1 as its symbols, as shown in 5.1.2 section. In on-off-keying (OOK), the peak-peak value of the voltage amplitude of the modulation signal, V_{PP}, is set to $V_{\pi \text{MZM}}$ to obtain the maximum and minimum light intensity states and these are used as symbols. In BPSK, the voltage amplitude V_{PP} is set to $2V_{\pi \text{MZM}}$. As shown in Fig. 4.29, this corresponds to one cycle of light intensity change, and although the intensities are equal for the two symbols, as mentioned above, in the individual phase modulators this corresponds to $V_{\pi \text{PM}}$, so they are in opposite phase to each other.

By applying this technique to the two quadrature components of the light wave, we can realize more complex optical signal generation, such as quadrature-phase-shift-keying (QPSK) and 16-level quadrature amplitude modulation (16QAM), etc. For more details, please refer to Chap. 5.

In addition to the push-pull operation, we consider various operation conditions, where the modulating signals are expressed by

$$v_1(t) = g(t) + h(t) \tag{4.42}$$
$$v_2(t) = -g(t) + h(t). \tag{4.43}$$

The optical output is given by

$$R = \frac{KE_0 e^{ih(t)}}{2} \left[e^{ig(t)} + e^{-ig(t)} \right] e^{i\omega_0 t}$$
$$= KE_0 e^{i\omega_0 t} \cos[g(t)] e^{ih(t)}, \tag{4.44}$$

where the amplitude $E(t)$ and phase $\Phi(t)$ are

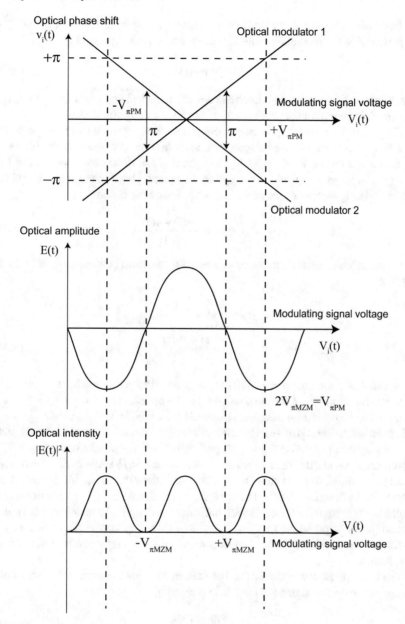

Fig. 4.29 Optical phase shift by two phase modulator embedded in an MZM, and optical amplitude and intensity of the MZM output with push-pull operation condition: $\alpha_0 = 0$

$$E(t) = E_0 \cos [g(t)] \tag{4.45}$$
$$\Phi(t) = h(t). \tag{4.46}$$

As both $h(t)$ and $g(t)$ describe induced optical phase by Pockels effect, $h(t)$ is proportional to the modulating-signal voltage and can be expressed by

$$h(t) = \alpha_0 g(t). \tag{4.47}$$

The proportional constant, α_0, describes deviation from the perfectly balanced push-pull operation condition, and is called the intrinsic chirp parameter.

The output light intensity is expressed by the same equation as in the push-pull operation, (4.39), and it can be seen that α_0 or $h(t)$ has no effect on the intensity. As mentioned above, $2g(t)$ refers to the optical phase difference between the two waveguides, whereas $h(t)$ is the average of the optical phase changes induced at the two modulating electrodes, as described in the following equation:

$$h(t) = \alpha_0 g(t) = \frac{v_1(t) + v_2(t)}{2}. \tag{4.48}$$

The intrinsic chirp parameter can be expressed by the ratio between $g(t)$ and $h(t)$, as follows:

$$\begin{aligned} \alpha_0 &= \frac{h(t)}{g(t)} \\ &= \frac{v_1(t) + v_2(t)}{v_1(t) - v_2(t)}. \end{aligned} \tag{4.49}$$

Figure 4.30 shows the phase shifts at the two optical phase modulators $v_1(t)$ and $v_2(t)$, and the amplitude $E(t)$ and phase $\Phi(t)$ of the optical output, when $\alpha_0 = 1/3$. As in Fig. 4.29, the horizontal axis is normalized by $V_{\pi\text{MZM}}$. In both cases of $\alpha_0 = 0$ and $1/3$, the modulation signal voltage can be switched from the on-state to the off-state by changing the modulation signal voltage from 0 to $+V_{\pi\text{MZM}}$. In the case of $\alpha_0 = 1/3$, the optical phase shifts at the two phase modulators are $4\pi/3$ and $-2\pi/3$. The average of the phase shifts, which is equal to $\pi/3$, corresponds to the phase shift in the optical output, Φ. As described below, the intrinsic chirp parameter α_0 is a quantity that is related to the chirp parameter α, which indicates the degree of parasitic optical phase change (parasitic phase modulation) associated with amplitude modulation. $\alpha_0 = 0$ corresponds to ideal amplitude modulation and $\alpha_0 \rightarrow \infty$ corresponds to pure phase modulation.

The chirp parameter is defined as the ratio of the phase change to the amplitude change in the optical signal [13] and is expressed by

$$\alpha = \frac{d\Phi}{dt} \bigg/ \frac{1}{E}\frac{dE}{dt}. \tag{4.50}$$

Substituting (4.45), (4.46), (4.47) to (4.50), we obtain

$$\alpha = -\alpha_0 \cot\left[g(t)\right], \tag{4.51}$$

where the parasitic phase modulation depends on $g(t)$ in general.

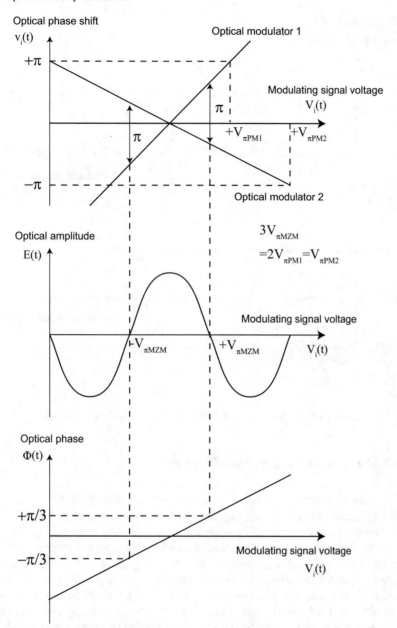

Fig. 4.30 Optical phase shift by two phase modulator embedded in an MZM, and optical amplitude and phase of the MZM output with $\alpha_0 = 1/3$

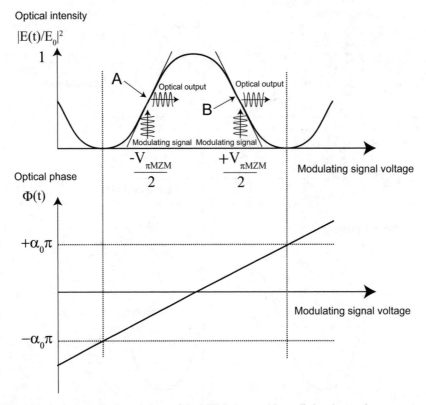

Fig. 4.31 Optical amplitude and phase of the MZM output with small signal operation

4.3.3 Bias Condition and Chirp Parameter

In intensity modulation, a point at which the optical intensity $|E(t)|^2$ equals half of the maximum value is commonly used as a bias point, where $g(t) = \mp\frac{\pi}{4}$. The bias condition can be achieved by applying a DC bias voltage of $\pm V_{\pi\mathrm{MZM}}/2$ with respect to a voltage level of an on-state or an off-state. As with common electrical amplifiers, it is necessary to properly set the average voltage of the modulating signal, which fluctuates alternately. This average voltage is called the DC bias or simply the bias of the modulator. The bias of the MZM corresponds to the average optical phase difference between the light waves from the two phase modulators. In digital communication systems, an OOK signal can be obtained by setting the amplitude of the modulation signal to $V_{\pi\mathrm{MZM}}/2$ as shown in the 5.1.1 section. This bias condition is commonly used also for analog communication systems, where high linearity is required in waveform transfer [14]. This bias condition is called quadrature bias in this book, since the phase difference between the optical outputs from the two phase modulators equals 90°. The biases at the points where the optical output intensity is maximum and minimum are called full bias and null bias, respectively.

Here, we consider the chirp parameter of the optical output from an MZM with the quadrature bias. The induced optical phase with a small signal modulation can be given by

$$g(t) = \mp \frac{\pi}{4} + a(t), \quad |a(t)|^2 \ll 1. \tag{4.52}$$

By referring to (4.51), the chirp parameter can be expressed by the intrinsic chirp parameter, as follows:

$$\alpha = \pm \alpha_0. \tag{4.53}$$

In other words, α_0 is an index showing the degree of parasitic phase modulation. By assuming small signal modulation ($|a|^2 \ll 1$), the intrinsic chirp parameter α_0 matches the chirp parameter α, under the quadrature bias condition commonly used in intensity modulation. For a small amplitude modulation signal with the bias set to point A $[-V_{\pi\mathrm{MZM}}/2]$ shown in Fig. 4.31, the optical output changes in phase and the chirp parameter $\alpha = \alpha_0$. On the other hand, at point B $[+V_{\pi\mathrm{MZM}}/2]$, the sign of the waveform is inverted where $\alpha = -\alpha_0$.

4.3.4 On-Off Extinction Ratio

When the optical phase difference between the two light waves from the two phase modulators is zero, as shown in Fig. 4.22, the optical signals are enhanced by interference at the output, while when the optical phase difference is 180°, they are converted to radiative higher-order mode light at the Y-junction for mixing, ideally resulting in zero output. However, in an actual modulator, the output does not become zero in the off-state, because there is a residual component due to the coupling of the higher-order radiative mode light to the guided fundamental mode light (crosstalk) and the intensity imbalance between the light waves from the two phase modulators, as shown in Fig. 4.32. The ratio of this residual component to the intensity of the on-state is called the on-off extinction ratio, or simply the extinction ratio (ER), and is an important index showing of the accuracy of the MZI.

By substituting the transmittance K_1 and K_2 defined by

$$K_1 = \left(1 + \frac{\eta}{2}\right) \frac{K}{2} \tag{4.54}$$

$$K_2 = \left(1 - \frac{\eta}{2}\right) \frac{K}{2}, \tag{4.55}$$

to (4.34), the optical output can be expressed by

$$
\begin{aligned}
R &= \frac{K e^{i\omega_0 t}}{2} \left[\left(1 + \frac{\eta}{2}\right) e^{iv_1(t)} + \left(1 - \frac{\eta}{2}\right) e^{iv_2(t)} \right] \\
&= K e^{ih(t)} \left[\cos g(t) + i \frac{\eta}{2} \sin g(t) \right].
\end{aligned} \tag{4.56}
$$

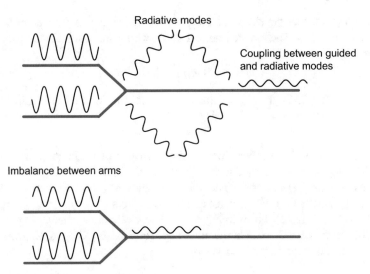

Fig. 4.32 Residual optical output in the off-state

η describes the imbalance of light intensity in the two arms in the MZ structure. It becomes zero when the two arms are perfectly balanced. In an actual modulator, η has a finite value due to structural asymmetries in the Y-junctions and waveguides due to manufacturing errors and optical loss imbalance in the two arms. Even for LN modulators, which are considered to have high accuracy, η would have values of about $\eta \sim 0.1$.

The second term of (4.56) is the component that degrades the extinction ratio, and if the desired component of the first term is taken as the real axis on the phasor, the phase is shifted by $\pi/2$ with respect to the real axis. The MZM controls the amplitude of the input light, and when the applied voltage is changed, the phase of the output light is constant and has the function of changing only the amplitude. It is interesting to note that the component that remains at the output light minimum, that is, in the off-state, is the component that is orthogonal on the phasor to the component to be controlled. While the real axis component can be made to match zero, the imaginary axis component takes its maximum value in the off-state, so the phase of the output light changes significantly in the near-off state.

Ratio of the maximum value of the output optical intensity $|R|^2$ to the minimum value

$$ER = \left(\frac{\eta^2}{4}\right)^{-1} \tag{4.57}$$

shows the ER, which is $-20 \log \eta/2$ in decibels. The induced optical phases $g(t)$ and $h(t)$ defined by (4.42) and (4.43) can be rewritten by

$$g(t) = \frac{v_1(t) - v_2(t)}{2} \tag{4.58}$$

$$h(t) = \frac{v_1(t) + v_2(t)}{2}. \tag{4.59}$$

$g(t)$ is half the optical phase difference between the two optical arms, which corresponds to the magnitude of the optical phase shift that should occur in each phase modulator in a balanced push-pull operation. $h(t)$ is the average of the phase changes in the two phase modulators and represents the phase change that occurs in the MZM optical output as shown in (4.48).

Light wave converted into radiative mode waves at the Y-junction in the MZM for optical output can be expressed by

$$R^* = K e^{ih(t)} \left[i \sin g(t) + \frac{\eta}{2} \cos g(t) \right]. \tag{4.60}$$

Suppose that the waveguide mode component R and the radiation mode component R^* may recombine from the vicinity of the combining section to the output port, the recombination process can be approximately expressed by a two-input, two-output directional coupler model, as follows:

$$\begin{bmatrix} R' \\ R^{*\prime} \end{bmatrix} = \begin{bmatrix} \cos \xi & -i \sin \xi \\ -i \sin \xi & \cos \xi \end{bmatrix} \begin{bmatrix} R \\ R^* \end{bmatrix}. \tag{4.61}$$

The optical output, which include the recombination process, can be given by

$$R' = K e^{ih(t)} \left[\cos \{g(t) - \xi\} + \frac{i\eta}{2} \sin \{g(t) - \xi\} \right]. \tag{4.62}$$

The optical output intensity normalized by K is

$$|R'/K|^2 = \cos^2 \{g(t) - \xi\} \left(1 - \frac{\eta^2}{4} \right) + \frac{\eta^2}{4}. \tag{4.63}$$

The intensity becomes a maximum when $g(t) - \xi = 0$. The maximum value is unity. The intensity becomes a minimum when $g(t) - \xi = \pi/2$, where the minimum value is $\eta^2/4$. η can be controlled by using an active trimming with a device as shown in Fig. 4.33, so that the imbalance can be precisely compensated [1]. In this case, the optical phase difference incorporates the effect of intermode coupling, and corresponds to the off-state when $\pi + 2\xi$.

Figure 4.34 shows a measured example of ER enhanced by the active trimming [1]. An integrated MZM with two arms consisting of a small MZM and an optical phase modulator connected in series can offer effective active trimming. The small MZMs correct the intensity imbalance of the main MZM. While ER of a common MZM is about $20 \sim 30$ ($\eta \sim 0.1$), ER of over 70 dB has been achieved by using the integrated MZM for the active trimming. This is close to the measurement limit of optical power meters and optical spectrum analyzers. A Y-junction with an electrode can be also used for imbalance trimming, to enhance the ER, where lightwave power splitting ratio at the Y-junction is controlled by the voltage applied to the electrode [15].

ER would be reduced due to coupling and scattering with higher-order mode components neglected in (4.61), but in the actual modulator, these effects are small as described above. Therefore, we can deduce that the approximate model with a 2×2 matrix accurately represents the characteristics of the interferometer in the MZM using LN.

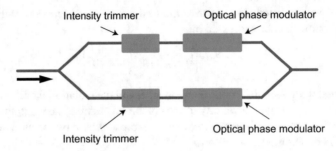

Fig. 4.33 MZM with active trimmers

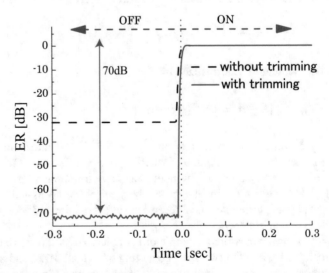

Fig. 4.34 ER enhancement by active trimming [1]

4.3.5 Chirp Parameter and Finite Extinction Ratio

Here, consider the case where the effects of the finite ER associated with η and the intrinsic chirp parameter described by α_0 exist simultaneously. Figure 4.35 shows a schematic of an MZM consisting of two optical phase modulators with balanced push-pull operation. The optical signal in each part of the modulator is shown in a phasor diagram. Each phase modulator generates phase modulation light that draws an arc in the complex plane, and by adding this vectorially, amplitude modulation in which the output light moves on the real axis can be obtained.

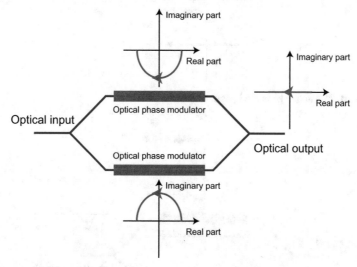

Fig. 4.35 Amplitude modulation by an MZM with balanced push-pull operation

Figure 4.36 shows the operation of the MZM in the case of $\eta \neq 0$. The component whose phase difference 90° with respect to the desired component causes the ER degradation in the MZM. The phase of the optical output deviates from 0 or 180° during the transition from the on-state to the off-state, and the deviation would be large when the state of the MZM is close to the off-state. That means a chirp component would be generated during the transition. Figure 4.37 shows the MZM optical output with a non-zero intrinsic chirp parameter $\alpha_0 \neq 0$, where the absolute values of optical phase shifts at the two phase modulators have some difference. Although the optical output moves on a trajectory off the real axis, it always passes through the origin if $\eta = 0$. Thus, α_0 does not affect the extinction ratio, as is clear from (4.56).

The ER of a typical MZM is about 20 dB and the chirp parameter α_0 is about $0.1 \sim 0.2$. The phasor diagram in Fig. 4.38 shows the trajectory of the optical output of the MZM when $\alpha_0, \eta \neq 0$. For simple modulation schemes, under the condition of $|\alpha_0|, |\eta| < 0.2$, the amplitude modulation has been considered to be sufficiently

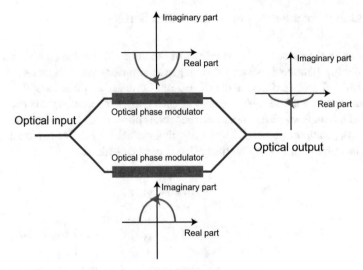

Fig. 4.36 MZM output with optical power imbalance

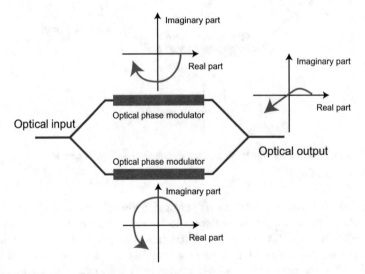

Fig. 4.37 MZM output with modulating-signal imbalance

accurate, but for complex modulation schemes such as 16-QAM and 256-QAM, there are issues such as distortion in the symbol positions on the phasor.

In the case of $\alpha_0 \neq 0$, the trajectory on the phasor can be changed by η. That implies that there is a possibility of improving the preciseness of the amplitude modulation by adjusting the ER through η. As mentioned in the previous section, it is relatively easy to change η by adjusting the optical intensity difference of the interferometer,

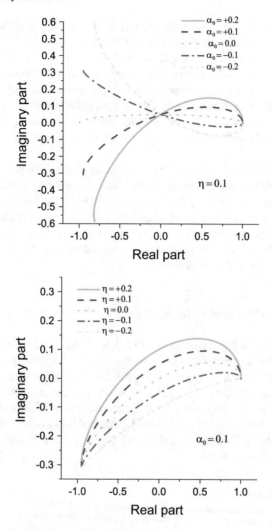

Fig. 4.38 MZM optical output with finite ER and chirp parameter

so that the method of suppressing the effect of α_0 caused by manufacturing error by adjusting η is considered to be effective.

4.3.6 Practical Implementation of Mach-Zehnder Modulators

The principle of operation of the MZM as an intensity modulator is to convert the phase difference $2g(t)$ into an intensity change using an interferometer, as shown in

(4.39). In order to improve the efficiency of the modulator, it is important to obtain a larger phase change $g(t)$ with a smaller modulation voltage.

Push-pull operation, in which optical phase change is applied in both of the two optical paths of the interferometer, is an effective way to improve the modulation efficiency. For this purpose, it is important that $v_1(t)$ and $v_2(t)$ in (4.42) and (4.43) have opposite signs and large values. In order to suppress the parasitic phase modulation (chirp), it is necessary to make the intrinsic chirp parameter α_0 close to zero as $v_1(t) \simeq -v_2(t)$ as shown in (4.51). This corresponds to minimizing $h(t)$ as defined in (4.48).

To keep the phase difference and minimize the intrinsic chirp parameter, there are two methods: one is to supply signals to two phase modulators separately through the driving circuit (circuit to supply modulation signals to the modulators), and the other is to realize the simultaneous application of electric field to two phase modulators with one electrode structure by devising the device structure. In this subsection, the methods for differential signal generation and various device structures for balanced push-pull operation will be explained.

4.3.6.1 Differential Signal Generation for Push-Pull Operation

Figure 4.39 shows a schematic of an MZM with two modulating-signal inputs. Two electrical signals are applied independently to the modulating electrodes of the two phase modulators. It is necessary to generate modulation signals $+V(t), -V(t)$, whose amplitudes are equal in magnitude and whose signs are inverted with respect to each other, in an external circuit. Such signals are called differential signals. Three methods are introduced here: the inverted output of a high-frequency amplifier, a hybrid coupler, and differential signal generation by phase difference generation due to differences in cable length.

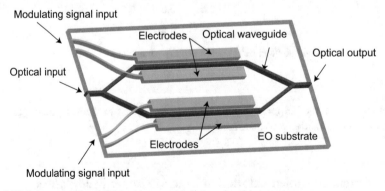

Fig. 4.39 Schematic of an MZM with two modulating-signal inputs

Differential Signal Generation Using Inverted Output of High-Frequency Amplifier

Figure 4.40 shows the configuration of differential signal generation using the inverting and non-inverting outputs of a high-frequency amplifier. From a common input, two outputs $+V(t)$, $-V(t)$ with their signs reversed are obtained. Transistors can offer function of inverting and amplifying waveforms. Inverting and non-inverting electrical output for signals in a wide frequency band can be obtained by using the functions of transistors in the amplifier. The signals can be amplified to an appropriate amplitude just before the modulator input. In binary digital modulation, it is effective to use a logic circuit such as a flip-flop to adjust the signal timing and waveform. On the other hand, there are issues such as variation of device characteristics inside the amplifier, output fluctuation due to temperature change, etc. In addition, it is necessary to improve the linearity of the amplification characteristics in order to support analog modulation and multilevel modulation.

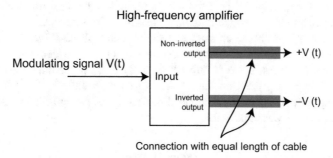

Fig. 4.40 MZM driving circuits based on differential signal generation by a high-frequency amplifier

Differential Signal Generation Using a Hybrid Coupler

Figure 4.41 shows the configuration of the drive circuit using a 180° hybrid coupler. In principle, a hybrid coupler can split an electrical signal into two without loss [16]. The configuration is similar to the case of using the inverting and non-inverting outputs of an amplifier, but since the operation of the hybrid coupler is basically linear, there is no problem of waveform degradation due to nonlinear distortion. On the other hand, due to the operating principle of the hybrid coupler, it is difficult to operate in a low frequency band, and the phase accuracy of the output signal tends to be poor in devices designed for a wideband operation.

Whether using an amplifier or a hybrid coupler, to achieve accurate modulation, it is important that the amplitudes of the two modulator inputs are equal and that the phase shift is exactly 180°, by balancing the cable length and adjusting the amplitude balance at the modulator inputs. In addition, the signal propagation distance and loss from the input ports to the electrodes inside the modulator also should be balanced.

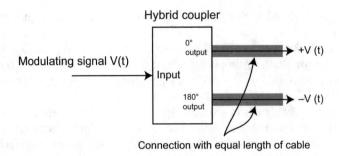

Fig. 4.41 MZM driving circuits based on differential signal generation by a hybrid coupler

Differential Signal Generation by Cable Length Difference

Figure 4.42 the configuration of the driving circuit with two different length cables. Two modulating signals of the same amplitude and phase are obtained at the power divider, and the propagation delay time difference is adjusted by using cables of different lengths, to apply an electrical signal with a phase difference of 180° to the modulator. The power divider can divide the signal in wideband from low to high-frequency regions, but it is difficult to apply it to broadband signals because the phase difference generation due to the cable length difference is 180° only at certain frequency components. Another disadvantage of power dividers is that they lose half of their power in principle.

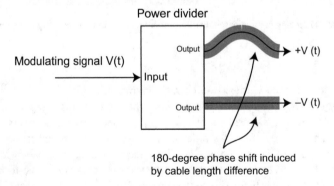

Fig. 4.42 MZM driving circuits with two different length cables

Comparison of Various Differential Signal Generation Methods

Table 4.1 shows the advantages and disadvantages of each of the above methods. Further, in the generation of the phase difference by the use of waveform inversion in amplifiers, the power consumption of the amplifier and the heat generated by the amplifier may become a problem. The hybrid coupler has a well-balanced performance in each item. Reference [16] shows an example of a wideband hybrid coupler that achieves 180° phase difference with a center frequency of 7 GHz and a bandwidth of 8 GHz. In the case of the 180° phase difference caused by the cable length difference, the bandwidth is only about 1.4 GHz even if the tolerance is less than ±10%. Since the length of the hybrid coupler depends on the wavelength of the target signal, devices that cover lower frequency ranges have the problem of increasing in size.

Table 4.1 Comparison of MZM driving methods

	Waveform inversion in amplifier	Hybrid coupler	Different length cables
High-frequency	Good	Good	Excellent
Wideband operation	Good	Fair	Bad
Linearity	Fair	Good	Good
Signal power	Excellent	Good	Fair
Stability	Fair	Good	Good

4.3.6.2 Device Structures for Push-Pull Operation

As described above, it is not easy to generate differential signals whose signs are inverted from each other. Here, we would like to introduce modulator structures that enables push-pull operation with a single modulation signal input.

Z-Cut MZM

In z-cut LN substrates, strong electric fields can be induced near electrodes edges, as shown in Fig. 4.12. By placing two optical waveguides near points A and B, effective push-pull operation can be achieved, where the induced optical phase shifts have opposite signs each other. The electric field intensity close the edge of the ground electrode is less than that of the signal electrode, where $h(t) \neq 0$, that is, the push-pull operation is not balanced. Although it is difficult to achieve balanced push-pull operation, this device structure is very useful for effective intensity modulation. The optical modulators with the cross section shown in Fig. 4.43a has been commonly used for intensity modulation, and is often referred to simply as a z-cut MZM.

The intrinsic chirp parameter α_0 is generally about 0.7, where the EO effect at the ground electrode side is about 1/6 compared to near the signal electrode edge. In optical fiber communication, mitigation of dispersion effect in optical fibers is one of important issues. Fiber dispersion degrades intensity modulation signals during transmission because the dispersion effect converts intensity modulation into phase modulation, and vice versa. In digital transmission systems using OOK, parasitic phase modulation would suppress the dispersion effect and improve the transmission performance. This is the reason why the z-cut LN modulators were widely used in practical optical fiber systems. Parasitic phase modulation would be converted into intensity modulation through the dispersion effect. At the receiver side, the intensity modulation signal converted from the parasitic modulation component would compensate degradation of the intensity modulation component sent from the modulator.

However, analog modulation, which requires multilevel modulation and highly accurate waveform transmission, which have recently been expanding, requires amplitude and intensity modulation close to the ideal [1, 17, 18]. Although it is possible in principle to balance the push-pull operation by shifting the optical waveguide on the signal electrode side from the point of maximum electric field strength, it reduces the modulation efficiency and leads to an increase in $V_{\pi MZM}$. Another issue is that near the edge, the change in electric field strength due to position change is large, and a small position shift leads to a change in modulation characteristics.

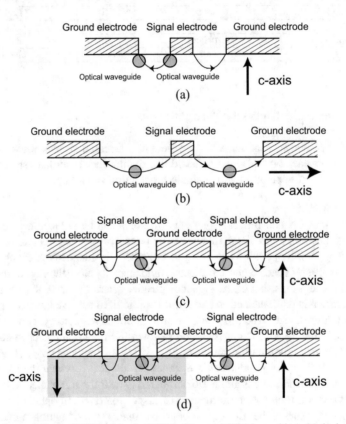

Fig. 4.43 Cross sections of MZMs. (**a**) z-cut MZM, (**b**) x-cut MZM (zero chirp), (**c**) dual-electrode MZM and (**d**) z-cut MZM (zero chirp)

X-cut MZM

In x-cut LN substrates, electric fields along the c-axis have maxima near the middle of the ground and signal electrodes, for example, points B and E in Fig. 4.13. The field has the same intensity at the point B and E. On the other hand, the direction of the electric field is opposite to each other at the two points. Thus, ideal well-balanced push-pull operation can be achieved by placing two optical waveguides at points B and E, as shown in Fig. 4.43b, where $v_1(t) \simeq -v_2(t)$, i.e., $h(t) \simeq 0$. This device structure is widely used in commercial modulators and is called an X-cut MZM.

It is also called a zero-chirp modulator or a low-chirp modulator because the intrinsic chirp parameter α_0 can be kept below 0.2. As shown in (4.51), a near-ideal amplitude modulation is obtained with almost zero parasitic phase modulation. A vector modulator that can handle complex modulation schemes has also been realized by combining several of these modulators.

Dual-Electrode MZM

An MZM with two separate electrodes, shown in Fig. 4.39, can be used for well-balanced push-pull operation. This modulator is often called a dual-electrode MZM, whose cross section is shown in Fig. 4.43c. Z-cut substrates with optical waveguides placed near edges of signal electrodes are commonly used to obtain large EO effect. This is the most advantageous structure in terms of reducing the half-wave voltage $V_{\pi \text{MZM}}$. An electric field is applied to the two optical waveguides through modulating-signal electrodes provided separately. By adjusting the signal amplitude and phase using an external drive circuit, precise amplitude modulation and single-sideband (SSB) modulation can be achieved, but as mentioned above, the external circuitry would be complicated.

Polarization-Reversed Structure z-Cut MZM

A z-cut zero-chirp modulator using polarization reversal has been proposed with the aim of achieving both balanced push-pull operation with a single modulation signal input as of an x-cut modulator and high efficiency modulation using a strong electric field near the signal electrode as of a z-cut modulator [19]. The cross sectional view is shown in Fig. 4.43d, which has two separate modulation electrodes and an optical waveguide near each signal electrode to obtain large EO effect. The basic structure is the same as that of the dual-electrode MZ modulator in Fig. 4.43c, but the feature is that the direction of the c-axis is reversed vertically in the left and right halves of the cross sectional view. The polarization direction can be reversed by applying a high voltage to a ferroelectric material such as LN or LT. Since one of the waveguides is in a crystalline region where the polarization is reversed, the direction of the optical phase change has the opposite sign even if the applied electric fields are in the same direction.

When two signals of the same sign and amplitude are generated using a signal divider and applied to the two modulating electrodes, a balanced push-pull operation is achieved. It is relatively easy to equally divide a high-frequency signal into two signals of the same sign and amplitude. By integrating the distribution circuit within the modulator as shown in Fig. 4.44, zero-chirp modulation is efficiently achieved with a single modulation signal input [19]. As with the x-cut modulator, no external circuitry is required to achieve near-ideal amplitude modulation of $\alpha_0 \simeq 0$. Note that actual device structures should be more complicated than shown in Fig. 4.44, to maintain impedance matching condition at the junction which feeds the electric signal to the two electrodes, and to suppress undesired reflection at the electrode ends.

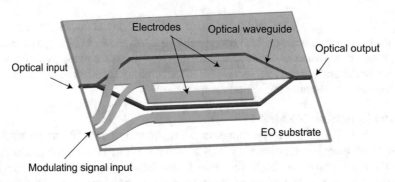

Fig. 4.44 Z-cut zero-chirp MZM with an electric signal divider

Figure 4.45 shows another configuration for low-chirp modulation with a polarization-reversed structure [20]. The first half of the EO substrate with respect to the signal propagation direction has the c-axis along the upward direction, while the c-axis in the second half is along the downward direction, as shown in Fig. 4.46. As described above, a z-cut MZM for push-pull operation has some imbalance due to difference between the electric field intensity under the signal electrode and that under the ground electrode as shown in Fig. 4.12. In the structure shown in Fig. 4.45, the absolute values of the induced optical phase shifts in the arms 1 and 2 can be equalized by shifting the locations of the signal and ground electrodes with respect to the optical waveguides of the MZM. The electric field amplitude induced by the modulating signal along the c-axis is negative in the arm 1 in the first half part as shown in Fig. 4.46a, where the waveguide of the arm 1 is placed close to the edge of the signal electrode. In the second half part, the optical waveguide of the arm 1 is under the edge of the ground electrode, so that the direction of the electric field is flipped with respect to that in the first half part. Due to the polarization-reversed structure, the amplitude of the electric field along the c-axis is negative, in other words, same as in the first half part. Thus, the induced optical phase can be effectively accumulated in the optical waveguide of the arm 1. Similarly, the induced electric field in the arm 2 is positive both in the first and second halves of the EO substrate. Due to the symmetry of the device structure, the intrinsic chirp parameter can be small or close to zero.

When we can neglect electric signal propagation loss on the electrode and velocity mismatching described in Sect. 4.2.4, the lengths of the first and second halves should be equal to each other to minimize the intrinsic chirp parameter. On the other hand, the modulation efficiency in the second half part would be less than in the first half, if the propagation loss or the velocity mismatching is not negligible. For low intrinsic chirp parameter with ideal push-pull operation, the lengths of the second half should be larger than that of the first half. Numerical calculation on the signal propagation loss and velocity mismatching is needed to design the polarization-reversed structure for low-chirp and effective modulation [20]. This structure can offer low half-wave voltage and low chirp with a simple electrode which is applicable to high-frequency operation. In addition, polarization-reversed structures can be used for modulators with various functions. MZMs with complex polarized domains can offer arbitrary frequency or impulse responses [21, 22].

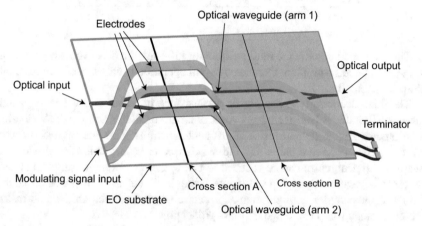

Fig. 4.45 Z-cut zero-chirp MZM with a polarization-reversed structure in the propagation direction

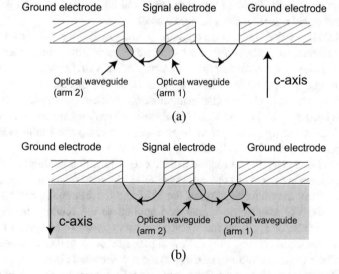

Fig. 4.46 Cross sections of Z-cut zero-chirp MZM with a polarization-reversed structure in the propagation direction. (**a**) cross section in the first half part, and (**b**) that in the second half part in Fig. 4.45

Comparison of Various MZ Modulators

Table 4.2 summarizes the features and characteristics of various MZM.

Table 4.2 Comparison of various MZMs

	Z-cut	X-cut	Dual-electrode	Z-cut zero-chirp
Low driving voltage	Good	Fair	Excellent	Good
Low chirp	Bad	Good	Excellent	Good
Ease of manufacture	Good	Good	Bad	Fair

The z-cut modulator is the most suitable for lowering the drive voltage (reducing $V_{\pi MZM}$), but it has a problem that the push-pull operation is unbalanced and residual chirp is not small.

The dual-electrode type has higher performance in terms of drive voltage and low chirp, but this is because it has two external drive circuits and the drive circuits have precise amplitude and phase adjustment functions. This means that if there are two external drive circuits and the drive circuits have precise amplitude and phase adjustment, chirp can be reduced to below the measurement limit. This can be used when extremely high performance is required, such as in advanced research. However, the complex configuration and precise adjustments required would make this technology still a challenge for mass production.

The z-cut zero-chirp modulator, as mentioned above, aims to achieve both high accuracy and simple configuration as an amplitude modulator of the x-cut modulator and low drive voltage of the z-cut modulator. However, the additional process of polarization inversion is required for fabrication, and the realization of an integrated and efficient signal distributor is a technical challenge.

Table 4.2 shows that the x-cut modulator is not suitable for reducing the drive voltage. However, as described in Sect. 4.3.7, due to recent improvement of the modulation efficiency by development of new device structures and progress of driving circuits, it is becoming less of a major issue in practical use [23]. On the other hand, with the spread of multilevel modulation, the demand for high modulation accuracy is becoming higher and higher, so the use of x-cut modulators is expanding. A typical intrinsic chirp parameter α_0 of an x-cut zero-chirp modulator is less than 0.2 and optical insertion loss is 2–10 dB.

An MZM has an electrode structure to efficiently transmit a modulating signal of high-frequency components to optical waveguides. As shown in Fig. 4.47, DC bias voltage to control the bias condition of the MZI and the fast changing modulating signal are mixed in an electric circuit and applied to the modulation electrode. The bias electrode for applying the voltage for bias control (bias voltage) and the modulation electrode for the modulation signal can be installed separately (see Fig. 4.48). This has the advantage of simplifying the electrical circuit. However, it is necessary to allocate a part of the finite device length, which is limited by the wafer size (about 3–5 in.), to the bias electrode, resulting in a short effective modulation electrode. Therefore, there is a trade-off between reducing the power required for modulation and simplifying the drive circuit, and this is one of the design elements that needs to be optimized.

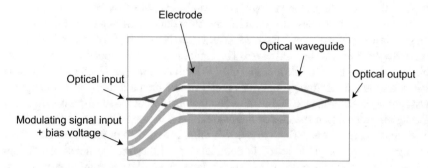

Fig. 4.47 MZM with a common electrode for modulating signal and bias voltage

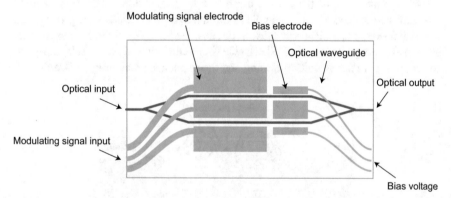

Fig. 4.48 MZM with separate electrodes for modulating signal and bias voltage

4.3.7 Advanced Device Structures

This section describes some state-of-the-art device structures for high-performance optical modulators. As described in Sect. 4.2.3, the modulation efficiency per unit length of the electrode is expressed by W defined by (4.18), which depends on the electrode structure and optical waveguide configuration. Modulation efficiency of each phase modulator embedded in an MZM can be described by Γ, which is defined by (4.19) and is proportional to the inverse of the half-wave voltage, as follows:

$$\Gamma = \frac{\pi}{V_{\pi_{PM}}} = -\frac{2\pi LW}{\lambda_0}. \tag{4.64}$$

The modulation efficiency denoted by Γ depends on the following three parameters: (1) the electrode length, (2) the intensity of electric field induced by the modulating signal, and (3) the EO coefficient of the substrate. The first item is directly described by L, while the effect of the last two items can be expressed by W.

To use long electrodes, the velocity and impedance matching condition should be satisfied as discussed in Sect. 4.2.4. The refractive index of LN for RF signals is much larger than that for lightwaves. The signal propagation speed largely depends on the refractive index of the material used to form the waveguide structure. Thus, an RF signal on the electrode propagates much slower than a lightwave on the optical waveguide, in general. One of the important factors in electrode designs is to increase the signal propagation speed. For lightwaves, the optical waveguide is formed by a Ti-diffused core whose refractive index is slightly larger than the substrate, where the lightwave propagation speed is dominated by the refractive index of LN. For RF signals, the signal propagation speed depends on the waveguide structure as well as on the refractive index of the material. The effective refractive index which is associated with the signal propagation speed depends on the average material refractive index weighted with the electric field distribution. Thus, a waveguide structure which leaks electric field distribution into air can enhance the propagation speed. For example, a CPW with thick electrodes, shown in Fig. 4.49, can increase the RF signal propagation speed [7, 8]. The CPW structure should be carefully designed to maintain both the velocity matching and the impedance matching conditions.

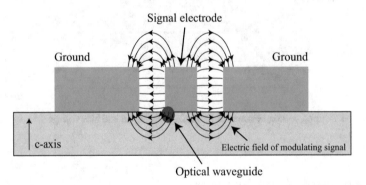

Fig. 4.49 Thick electrode for velocity matching

To increase the intensity of the electric field, a thin layer whose refractive index is higher than in other layers can be used as a substrate for the optical waveguide. The electric field is concentrated in an area where the refractive index is high. As shown in Fig. 4.50, by using a thin LN layer, the electric field of the modulating signal can be focused into the layer where the optical waveguide is embedded.

The thin LN layer structure is also useful for high-speed operation. In principle, a substrate would support a waveguide mode as with a rectangular waveguide. This undesired waveguide mode couples with the desired CPW mode, so that the frequency response would be degraded by interference between the two modes. The waveguide mode has a cutoff frequency, which is inversely proportional to the size of the waveguide structure. Thus, when the size is small enough, the modulating-signal frequency can be less than the cutoff frequency. The LN substrate would support a waveguide mode with the lowest cutoff frequency, because of the large refractive index. That means that the waveguide mode supported by the LN substrate would couple with the desired CPW mode. Thus, if the LN layer is thin enough to let the cutoff frequency higher than the frequency of the modulating signal, we can suppress the undesired interference which causes degradation of the frequency response. By using the thin LN layer structure, as shown in Fig. 4.51, we can obtain a flat frequency response up to 50 GHz [23].

A frequency response of a modulator measured by sweeping the frequency of the modulating signal is called an electro-optic (EO) frequency response. Figure 4.52 shows EO frequency responses of modulators based on a thin LN substrate and a conventional LN substrate [24]. The thickness of the thin LN substrate is less than 100 μm, while that of the conventional LN substrate is larger than 500 μm. The EO frequency response of the thin LN modulator is evaluated from optical sideband components (see Fig. 8.2). On the other hand, that of the conventional LN modulator for the frequency range less than 67 GHz is measured by using electric domain frequency sweep where the optical signal is converted into an electric signal through a photodiode (see Fig. 8.9). The frequency response for the frequency range higher than 67 GHz is measured from optical sideband components as for the thin LN modulator. The thin LN modulator has a flat frequency response over 100 GHz, while the performance of the modulator with the conventional LN substrate is largely degraded in a frequency range over 70 GHz. The EO responses are normalized as 0 dB at 10 GHz. Details on frequency response measurement methods are described in Sect. 8.3.

Ridge structures are also useful to enhance the electric field [25]. The thin LN and ridge structures are also can be used to optimize the effective refractive index and the characteristic impedance for the velocity and impedance matching. Ridge or other structures for optical waveguides are also useful to enhance the interaction between the RF modulating signal and the lightwave. A thin LN film with a silicon substrate can provide effective optical modulation, where the size of the optical waveguide is much smaller than in a conventional LN modulator with a Ti-diffused core [26–28]. In a device structure described in [28], the optical signal propagates along a silicon waveguide placed close to the LN film.

As described in Sect. 3.2.3, semiconductor materials can be also used for EO modulation. Recently, various InP-based modulators have been developed for high-speed operation with compact device sizes. An InP modulator using a multi-quantum-well (MQW) layer as an optical waveguide core, can offer effective optical modulation, where the refractive index change by the EO effect can be enhance by QCSE. Figure 4.53 shows a schematic of a high-speed InP modulator with a low-electrical loss electrode consisting of a backborn travelling-wave electrode and small T-shape electrodes. We can design RF signal propagation speed and loss as well as characteristic impedance, by adjusting the structure of the backborn and T-shape electrodes. Thus, the RF signal propagation speed can be matched with that of the lightwave to obtain

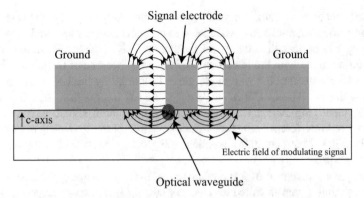

Fig. 4.50 Thick electrode for velocity matching and thin LN substrate for electric field concentration

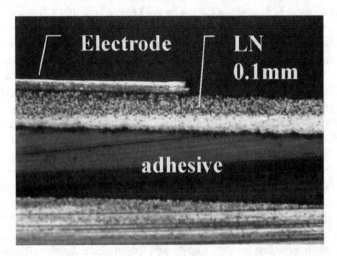

Fig. 4.51 Cross section of an LN modulator with a thin LN layer [23]

the velocity matching condition, without losing the impedance matching condition. Figure 4.54 shows an EO frequency response of the modulator reported in Ref. [29]. The EO frequency response is measured by a setup shown in Fig. 8.9. The half-wave voltage was 1.5 V at DC, and 3-dB bandwidth was about 80 GHz. The optical loss in the modulator was 8.5 dB. The half-wave voltage of an LN modulator with conventional Ti-diffused optical waveguides is typically 2–3 V, where the 3-dB bandwidth is not wider than 50 GHz. However, the optical loss is typically less than 5 dB. In short, InP-based devices can offer wideband frequency response, while conventional LN device structures can provide low optical loss.

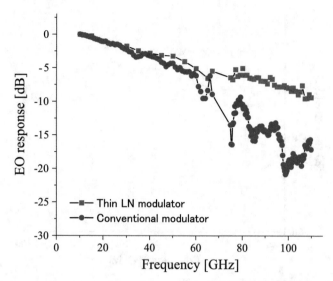

Fig. 4.52 EO responses of a thin LN modulator and a conventional LN modulator [24]

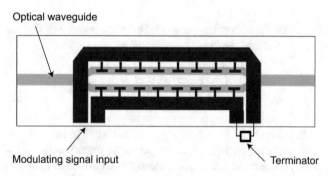

Fig. 4.53 Schematic of an InP modulator [29]

4.4 Vector Modulation

Amplitude and phase of the optical output can be expressed by a point on a complex plane. As shown in Fig. 4.55, an arbitrary point on the complex plane is generally represented using Cartesian coordinates (x, y) or polar coordinates (r, θ), as follows:

$$
\begin{aligned}
z &= x + \mathrm{i}y \\
&= r\mathrm{e}^{\mathrm{i}\theta},
\end{aligned}
\tag{4.65}
$$

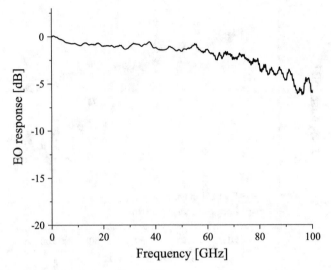

Fig. 4.54 EO response of an InP modulator [29]

where z denotes a complex number associated with the point. The Cartesian coordinates can be converted into the polar coordinates by

$$r = \sqrt{x^2 + y^2} \qquad (4.66)$$

$$\theta = \tan^{-1} y/x. \qquad (4.67)$$

A modulation method which controls the amplitudes and phase two-dimensionally is called vector modulation, since it corresponds to manipulating a vector in the complex plane. This section describes quadrature amplitude modulation (QAM), which controls the amplitudes of the real and imaginary components x and y on the complex plane independently, and amplitude phase modulation (APM), which controls the amplitude magnitude r and phase θ, respectively. These are realized by combinations of several modulators.

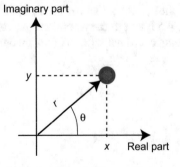

Fig. 4.55 Light wave described by a complex number

4.4.1 Quadrature Amplitude Modulation

An MZM can generate lightwave with arbitrary amplitude as described in (4.37). Figure 4.56 shows the amplitude modulation in the complex plane, where the optical output moves on the real axis according to the modulation signal. The amplitude component of the imaginary axis can be also controlled in the same way as shown in Fig. 4.57. Light waves with arbitrary states on the complex plane can be generated by combining these two components on the real and imaginary axes. This modulation method of independently controlling the amplitudes of the real and imaginary components to obtain various modulation signals is called quadrature amplitude modulation (QAM).

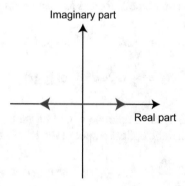

Fig. 4.56 Amplitude modulation for real part

A dual-parallel Mach-Zehnder modulator (DPMZM), which integrates two MZMs as shown in Fig. 4.58, can be used to control light waves in a vector manner. The optical input is divided into two parts by the Y-junction shown in Fig. 4.27 and led to the two MZMs. According to (4.36), optical outputs from the two MZMs can

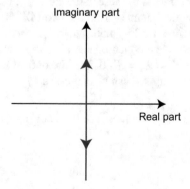

Fig. 4.57 Amplitude modulation for imaginary part

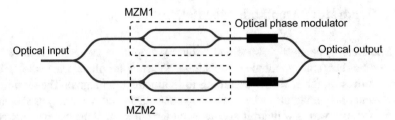

Fig. 4.58 Schematic of a dual-parallel MZM, where phase difference between the real and imaginary parts is controlled by an optical phase modulator

be controlled by $g_1(t)$ and $g_2(t)$ which are modulating signals of the MZM1 and MZM2 in Fig. 4.58. The two outputs, R_1 and R_2, can be expressed by

$$R_1 = \frac{1}{\sqrt{2}} K_1 E_0 e^{i\omega_0 t} \cos\left[g_1(t)\right] \tag{4.68}$$

$$R_2 = \frac{1}{\sqrt{2}} K_2 E_0 e^{i\omega_0 t} \cos\left[g_2(t)\right]. \tag{4.69}$$

As described in Sect. 4.3.1, the amplitude would be multiplied by $1/\sqrt{2}$ at the Y-junction for mixing, so that the optical output of the DPMZM can be given by

$$R = \frac{E_0 e^{i\omega_0 t}}{2} \left[K_1 \cos\left[g_1(t)\right] + e^{i\phi_p} K_2 \cos\left[g_2(t)\right]\right]. \tag{4.70}$$

Here, the initial phase of the optical input is set so that the phase of the optical output from MZM1 is zero at the output port of MZM1 at time $t = 0$. As we have discussed so far, the phase of the input light varies due to fiber fluctuations, etc., and the absolute value of the phase of the optical input would not have large impact on common transmission systems. The initial phase can be set as needed to simplify the mathematical expressions. While K_1 and K_2 describe loss in the optical waveguides of MZM1 and MZM2, the deviation from the ideal amplitude change $1/\sqrt{2}$ at the Y-junctions can also be expressed by these parameters. ϕ_p, which is the optical phase difference between the optical outputs from the two MZMs, can be adjusted by the optical phase modulator shown in Fig. 4.58. When $\phi_p = \pi/2$, the optical signal from MZM2 becomes an amplitude modulation signal on the imaginary axis as shown in Fig. 4.57. Assuming that $K_1 = K_2 = K$, that is, the optical loss in the two MZMs is well-balanced, the optical output R can be expressed by

$$R = \frac{E_0 K e^{i\omega_0 t}}{2} \left[\cos\left[g_1(t)\right] + i\cos\left[g_2(t)\right]\right]. \tag{4.71}$$

By using

$$g_1(t) = \cos^{-1} X(t) \tag{4.72}$$

$$g_2(t) = \cos^{-1} Y(t), \tag{4.73}$$

we get

$$
\begin{aligned}
R &= \frac{E_0 K e^{i\omega_0 t}}{2}[X(t) + iY(t)] \\
&= \frac{E_0 K e^{i\omega_0 t}}{2}\sqrt{X(t)^2 + Y(t)^2}e^{i\tan^{-1}\frac{Y(t)}{X(t)}}.
\end{aligned}
\tag{4.74}
$$

By referring to (4.3), the amplitude change and phase shift by this modulation method can be expressed by

$$E(t) = \frac{KE_0}{2}\sqrt{X(t)^2 + Y(t)^2} \tag{4.75}$$

$$\Phi(t) = \tan^{-1}\frac{Y(t)}{X(t)}. \tag{4.76}$$

Here the absolute values of the real and imaginary components, $|X(t)|$ and $|Y(t)|$ are less than unity. The optical output whose state is described as a point in the phasor diagram can be freely controlled within the range shown in Fig. 4.59, by changing the modulating signal $g_1(t)$ and $g_2(t)$. However, even assuming an ideal state ($K = 1$) where the optical loss is negligible, the output light amplitude is limited to $E_0/\sqrt{2}$ at the maximum. The maximum can be achieved when $X, Y = \pm 1$, which corresponds to the points indicated by A in Fig. 4.59. At points indicated by B, where $X = 1, Y = 0$, etc., the output optical amplitude is $E_0/2$. The energy conversion efficiency from input to output is 50% at point A and 25% at point B. More than half of the energy is lost as radiation in the Y-junctions for combining. As described in the Sect. 5.2.2, a multilevel QAM signal can be obtained by using the points where X, Y are equally divided in the range of $+1$ to -1 as symbols. Figure 4.60 shows symbols of a 16-level QAM signal, which can convey 4 bits of information in one time slot of the modulation signal by using 16 points of $X, Y = \pm 1$ or $\pm 1/3$ as symbols. A diagram showing the arrangement of symbols in the complex plane is called a constellation map or simply a constellation.

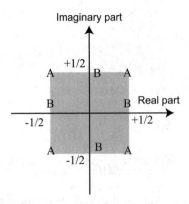

Fig. 4.59 Range of light wave states that can be generated by QAM

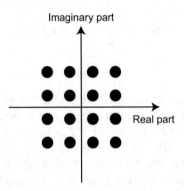

Fig. 4.60 Constellation map for 16QAM

4.4.2 Amplitude Phase Modulation

An optical signal can be expressed by product of the amplitude $E(t)$ and the element $e^{i\Phi(t)}$, which represents the phase rotation, as shown in (4.3). This means that an arbitrary light wave state can be obtained by combining amplitude and phase changes, as shown in Fig. 4.61, and vector modulation can be realized by combining phase and amplitude modulation in series. In principle, the same result can be obtained by interchanging the phase and amplitude modulation. Such a vector modulation by combining amplitude and phase modulation is called amplitude phase modulation (APM).

Figure 4.62 shows the configuration in which the amplitude is controlled by the MZM and the phase is rotated by the optical phase modulator.

By using (4.20) and (4.36), a mathematical expression for the optical output R can be obtained as follows:

$$R = K E_0 e^{i\omega_0 t + iv(t)} \cos\left[g(t)\right], \tag{4.77}$$

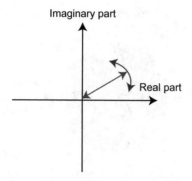

Fig. 4.61 Vector modulation by combination of amplitude modulation and phase modulation

where $g(t)$ and $v(t)$ are, respectively, the modulation signals applied to the MZM and the optical phase modulator. K denotes the overall optical loss in the MZM and phase modulator. By referring (4.3) and using

$$g(t) = \cos^{-1} X(t) \tag{4.78}$$

$$v(t) = Y(t), \tag{4.79}$$

we get

$$E(t) = K E_0 X(t) \tag{4.80}$$

$$\Phi(t) = Y(t), \tag{4.81}$$

which describes the function of APM. By changing $X(t)$ and $Y(t)$, we can generate any optical signals whose state are in the area shown in Fig. 4.63, where $0 \le X(t) \le 1$ and $0 \le Y(t) < 2\pi$.

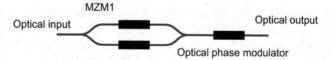

Fig. 4.62 MZM and phase modulator connected in series

Figure 4.64 shows a constellation map of the APM signal which consists of two amplitude levels with $X(t)$ and eight phase levels with $Y(t)$ (equally spaced by 45°) for a total of 16 states. As with 16QAM in Fig. 4.60, 16APM can convey 4 bits of information in one time slot of the modulation signal.

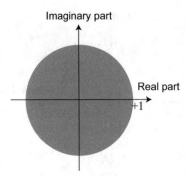

Fig. 4.63 Range of light wave states that can be generated by APM

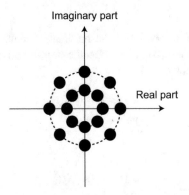

Fig. 4.64 Constellation map for 16APM

4.4.3 Comparison of Quadrature Amplitude Modulation and Amplitude Phase Modulation

Since QAM directly controls the amplitude of the real and imaginary parts of the modulating signal and complex numbers, as shown in (4.72), (4.73), and (4.74), so that QAM can be recognized as a physical implementation of manipulation of x, y in the Cartesian coordinate representation. On the other hand, APM can be said to be a physical realization of the manipulation of r and θ in the polar coordinate representation, since APM signal directly controls the magnitude and phase of the optical signal, as shown in (4.80) and (4.81).

Optical power conversion efficiency of the QAM using the DPMZM is at most 50%, due to intrinsic loss at Y-junctions. On the other hand, the conversion efficiency in APM can be larger than 50%, where the range of the optical signal in APM shown in Fig. 4.63 is much wider than in QAM shown in Fig. 4.59. When we neglect the loss in the modulators, that is, ($K = 1$), the maximum optical conversion efficiency of the APM is 100%. If the output of the MZM is set to the maximum, i.e., $X(t) = 1$, then the input light is converted to output light without loss. However, when the amplitude and phase are controlled separately in APM, distance between symbols with small amplitude becomes small as shown Fig. 4.64, where signal errors are likely to occur due to changes in the lightwave state caused by noise or distortion. In order to obtain an optical signal with equally spaced symbols as shown in Fig. 4.60, complex control is required for amplitude $X(t)$ and phase $Y(t)$. In the case of 16-QAM, it is necessary to vary the amplitude in 3 ways and the phase in 12 ways with unequal spacing. In electrical circuits that generate modulation signals, there is a trade-off between control precision and bandwidth. In high-speed transmission systems, it is not easy to generate multilevel QAM signals with evenly spaced symbols in the complex plane using APM. Thus, vector modulation using a DPMZM is often used despite its low conversion efficiency.

4.5 Stability of Optical Modulation

As mentioned above, various modulation schemes, such as amplitude modulation by an MZM and QAM by a DPMZM are realized by combining multiple phase modulators. The optical output would be affected by undesired phase fluctuation at each phase modulator. In this section, we discuss impact of such phase fluctuation on modulator function, and configurations which offer stable operation. The output optical phase of an optical phase modulator has large excursion due to light source fluctuation, temperature fluctuation of the fiber, mechanical vibration, etc. In addition, DC drift effect also induces optical phase shift.

As shown in (4.15), the initial phase at $t = 0$ can be set to zero. The phase oscillation in time domain is described by $e^{i\omega_0 t}$. This simple mathematical expression is useful to describe basic functions of modulators, however, optical phase, Φ_0, has large fluctuation. In actual communication systems shown in Fig. 4.65, optical phase is always fluctuating due to fluctuations in the light source and optical fiber. The phase change caused by the fluctuation is larger than the optical phase changes induced by optical modulation.

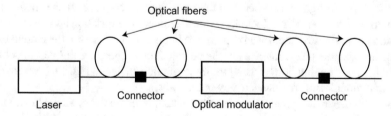

Fig. 4.65 Optical fibers connecting a laser and a modulator

4.5.1 Fluctuations in Lasers and Fibers

In the MZM, the outputs of two optical phase modulators are combined to achieve optical interference for amplitude control, where change of phase difference between in the optical paths is converted into intensity changes. Thus, the fluctuation in the phase difference has a significant impact on transmission systems that use optical intensity as information.

4.5.1.1 Fluctuations in Lasers

The configuration shown in Fig. 4.66 is commonly used in wavelength multiplexing transmission systems where light from lasers of different wavelengths is modulated individually and multiple channels are installed in the same fiber. Since wavelength-division multiplexing transmission systems do not use optical interference between different wavelength channels, random phase fluctuation in each laser has no effect on transmission performance.

If the wavelengths of the two lasers are matched in this configuration, the output of the combiner, which mixes the two optical signals, can be modulated in intensity by optical interference, in principle. However, it is rather difficult to stably keep the optical phase difference between the two optical signals generated by the two modulators, due to phase fluctuations of the laser light. In addition, it is very difficult to strictly match the wavelengths of the two independent light sources.

4.5.2 Fluctuations of Fiber Lengths

Impact of the phase or frequency excursion of lasers can be suppressed by using a common laser light source, which is bifurcated into two branches and phase modulated by two separate optical modulators, as shown in Fig. 4.67. Even if the wavelength of the laser fluctuates, there is no shift in the optical frequency difference between the two light waves because the fluctuation is common at the modulator input point. However, phase fluctuation due to temperature variation and mechanical vibration of the optical fibers used to guide the lights from the lasers to the modulators would have impact on the intensity of the optical output. Thus, it is difficult to obtain stable intensity modulation by using the configuration shown in Fig. 4.67, where the laser and modulators are connected by optical fibers.

On the other hand, optical path length fluctuation can be very small in an integrated MZM as shown in Fig. 4.39. Even if the optical path length itself changes with temperature, the changes are almost the same in the two optical paths. In addition, since the input light source is common, the influence of the above optical fiber fluctuation and phase fluctuation of the input light can be suppressed by the integration.

QAM described in Sect. 4.4.1 can be offered by two individual amplitude modulators which are connected by optical fibers, in principle. However, due to fiber length fluctuation, it is difficult to maintain the orthogonal relationship shown in Figs. 4.56 and 4.57 between the amplitude modulated signals obtained by the two modulators. Although control methods have been developed to stabilize the phase relationship between two light waves, it is necessary to detect the phase difference and realize feedback control, which poses problems such as configuration complexity and ensuring long-term stability. Integrated DPMZMs have been realized recently to offer stable vector modulation [1, 30–32]. In the integrated MZM for intensity modulation, the two light waves are on the same line on the phasor diagram to obtain stable interference, whereas in the integrated DPMZM for QAM, the two light waves are kept 90° out of phase with each other to achieve independent amplitude control without interference. For intensity modulation, the optical interference effect should be maximized by stabilizing the phase difference. On the other hand, the interference effect should be suppressed for QAM. However, these are based on the same technique in that it stabilizes the relative optical phase difference, by integration and use of a common laser.

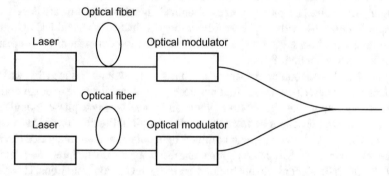

Fig. 4.66 Configuration with two laser sources for two modulators

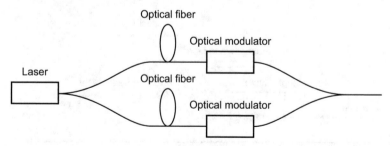

Fig. 4.67 Configuration with one laser source for two modulators

4.5.3 Suppression of Phase Fluctuation

In addition to the fluctuations in the light source (including fluctuations in the fiber connecting the light source to the modulator) and fluctuations in the optical path length of the optical fiber used to connect each modulator, there are also fluctuations in the optical path length of the modulator. This is due to DC drift in which the optical phase difference slightly shifts over a long period of time (even on the order of years or more). DC drift is caused by electric field induced by DC voltage applied to the electrode.

Here, we consider various types of fluctuations caused by laser instability, fiber vibration, temperature change and DC drift. The frequency component of each variation is limited to a specific range. Light source fluctuations are mainly in the range of a few MHz or less for common semiconductor lasers, and in the range of a few kHz or less for light sources with external resonators that offers narrow linewidth light waves [33, 34]. The bandwidth of the fluctuating component of the laser is generally referred to as the linewidth, as described in Sect. 4.2.2. Optical path length fluctuations due to temperature and mechanical fluctuations of the fiber are limited to lower frequency components. DC drift causes even slower fluctuations.

Figure 4.68 shows frequency ranges of optical signals whose bit rate is between 10 and 100 Gb/s, linewidths of lasers, fiber lengths fluctuations, and DC drift. For narrow linewidth lasers, the bandwidth of the fluctuation component is less than a few kHz or a few hundred Hz.

Table 4.3 shows counter measures for phase fluctuations induced by various mechanism. As described above, it is not necessary to stabilize the average optical phase at the output, but the relative phase difference between phase modulators connected in parallel should be fixed for stable operation. Fluctuations in lasers and fibers can be suppressed by using a common light source and an integrated device. On the other hand, DC drift causes optical phase change inside the integrated device. Thus, the DC drift effect would not be suppressed by device configurations. However, since DC drift is a very slow phenomenon, a common feedback control system can be used for automatic bias control which tracks the desired bias point [9, 35].

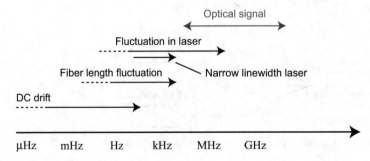

Fig. 4.68 Fluctuations caused by various factors

Table 4.3 Sources of phase fluctuations and suppression methods

	DC drift	Fiber length fluctuation	Laser fluctuation
Frequency range	a few years^{-1} ~ a few 10 Hz	~100 kHz	a few 10 MHz
Suppression methods	Feedback control	Integration	Common light source

In these cases, the absolute phase at the input and output ports of the modulator is less important. On the contrary, a configuration in which the absolute phase affects the overall performance or function will have stability issues in practical use. Since direct detection of intensity by a photodetector does not respond to changes in absolute phase, stable operation is possible as long as the relative phase difference, which affects the intensity change, is constant.

On the other hand, differential phase detection utilize phase difference between time slots adjacent to each other on the time axis, as symbols. If we use a narrow linewidth laser as a source, the optical signal changes faster than the laser fluctuations. Thus, stable demodulation can be achieved, where the effects of fluctuations can be suppressed between neighboring symbols, as shown in Fig. 4.68.

In digital coherent communication, which has become increasingly popular in recent years, optical phases are used as symbols for data transmission [36, 37]. Although narrow linewidth laser light sources are important for performance improvement [33, 34, 38], the optical phase at the transmitter side should be estimated and tracked by digital signal processing [39], where the absolute phase is not stabilized at the modulator input and output in these systems. As shown in Fig. 4.68, frequency ranges of signals for data transmission and fluctuations in laser overlap each other. Thus, we should rely on digital signal processing to suppress the optical phase change caused by optical path fluctuation and light source fluctuation. This configuration is called digital coherent transmission. It requires large computational resources, however, is commonly used in high-speed transmission systems. This can be attributed to a combination of three factors: (1) the fluctuations are limited to a certain frequency band due to the improved performance of the light sources, (2) the modulation speed has been improved and the frequency band of the optical signal is in a higher region, and (3) the capability of digital signal processing has been improved.

Problems

4.1 Calculate the refractive index change induced by the EO effect in an LN substrate, where the electric field of the modulating signal and that of the lightwave are parallel to the c-axis. For simplicity, assume that the electric field of the modulating signal is uniform between the signal and ground electrodes, where the electrode separation is 100 μm, and the voltage of the modulating signal is 10.0 V.

4.2 Calculate propagation delay difference between the extraordinary and ordinary rays, in an LN modulator whose length L is 5.00 cm.

4.3 Derive (4.23), and calculate the required signal power for the full-swing operation, as shown in Fig. 4.8, with an optical phase modulator whose half-wave voltage is 5 V.

4.4 Consider signal propagation speed difference as shown in Fig. 4.17. The relative permittivity of LN for electric signals polarized along the c-axis is

$$\varepsilon_{\mathrm{me}} = 28.0.$$

Assume that the effective refractive index of the modulating signal on the electrode is equal to the square root of the relative permittivity, that is,

$$n_{\mathrm{m}} = \sqrt{\varepsilon_{\mathrm{me}}}.$$

Calculate the cutoff frequency of a modulator with a 2-cm long electrode ($L = 2.00$ cm). The electric field of the lightwave is also parallel to the c-axis.

4.5 Derive (4.51).

4.6 Calculate the light intensity imbalance defined by (4.56) for an MZM whose ER is 60 dB.

References

1. T. Kawanishi, T. Sakamoto, M. Izutsu, High-speed control of lightwave amplitude, phase, and frequency by use of electrooptic effect. IEEE J. Sel. Topics Quant. Electron. **13**(1), 79–91 (2007)
2. T. Kawanishi, Integrated Mach-Zehnder interferometer-based modulators for advanced modulation formats, In *High Spectral Density Optical Communication Technologies, Optical and Fiber Communications Reports 6*, ed. by M. Nakazawa, K. Kikuchi, T.Miyazaki (Wiley, Hoboken, 2010)
3. T. Kawanishi, High-speed optical communications using advanced modulation formats, in *Wiley Encyclopedia of Electrical and Electronics Engineering*, ed. by J.G. Webster (Wiley, Hoboken, 2016)
4. M. Izutsu, Y. Yamane, T. Sueta, Broad-band traveling-wave modulator using a LiNbO$_3$ optical waveguide. IEEE J. Quant. Electron. **13**(4), 287–290 (1977)
5. R.C. Alferness, Waveguide electrooptic modulators. IEEE Trans. Microw. Theory Tech. **30**(8), 1121–1137 (1982)
6. G.L. Li, P.K.L. Yu, Optical intensity modulators for digital and analog applications. J. Lightwave Technol. **21**(9), 2010–2030 (2003)
7. K. Noguchi, H. Miyazawa, O. Mitomi, 75 GHz broadband Ti : LiNbO$_3$ optical modulator with ridge structure. Electron. Lett. **30**(12), 949–950 (1994)

8. K. Noguchi, O. Mitomi, H. Miyazawa, Millimeter-wave Ti : LiNbO$_3$ optical modulators. J. Lightwave Technol. **16**(4), 615–619 (1998)

9. A. Chen, E.J. Murphy, (eds.), *Broadband Optical Modulators* (CRC Press, Boca Raton, 2012)

10. L. Zehnder, Ein neuer interferenzrefraktor. Z. Instrum. **11**, 275–285 (1891)

11. L. Mach, über einen Interferenzrefraktor. Z. Instrum. **12**, 89–93 (1892)

12. M. Izutsu, A. Enokihara, T. Sueta, Optical-waveguide hybrid coupler. Opt. Lett. **7**(11), 549–551 (1982)

13. F. Koyama, K. Iga, Frequency chirping in external modulators. J. Lightwave Technol. **6**(1), 87–93 (1988)

14. T. Kawanishi, K. Kogo, S. Oikawa, M. Izutsu, Direct measurement of chirp parameters of high-speed Mach-Zehnder-type optical modulators. Opt. Commun. **195**(5–6), 399–404 (2001)

15. Y. Yamaguchi, A. Kanno, T. Kawanishi, M. Izutsu, H. Nakajima, Precise optical modulation using extinction-ratio and chirp tunable single-drive Mach-Zehnder modulator. J. Lightwave Technol. **35**(21), 4781–4788 (2017)

16. M.E. Bialkowski, Y. Wang, Wideband microstrip 180 ° hybrid utilizing ground slots. IEEE Microw. Wireless Components Lett. **20**(9), 495–497 (2010)

17. T. Kawanishi, *Wired and Wireless Seamless Access Systems for Public Infrastructure* (Artech House, 2020)

18. T. Kawanishi, A. Kanno, H.S.C. Freire, Wired and wireless links to bridge networks: seamlessly connecting radio and optical technologies for 5G networks. IEEE Microw. Mag. **19**(3), 102–111 (2018)

19. S. Oikawa, F. Yamamoto, J. Ichikawa, S. Kurimura, K. Kitamura, Zero-chirp broadband Z-cut Ti:LiNbO$_3$ optical modulator using polarization reversal and branch electrode. J. Lightwave Technol. **23**(9), 2756–2760 (2005)

20. N. Courjal, H. Porte, J. Hauden, P. Mollier, N. Grossard, Modeling and optimization of low chirp Ti : LiNbO$_3$ Mach-Zehnder modulators with an inverted ferroelectric domain section. J. Lightwave Technol. **22**(5), 1338–1343 (2004)

21. D. Janner, M. Belmonte, V. Pruneri, Tailoring the electrooptic response and improving the performance of integrated Ti : LiNbO$_3$ modulators by domain engineering. J. Lightwave Technol. **25**(9), 2402–2409 (2007)

22. H. Murata, Y. Okamura, High-speed signal processing utilizing polarization-reversed electro-optic devices. J. Lightwave Technol. **32**(20), 3403–3410 (2014)

23. T. Kawanishi, T. Sakamoto, A. Chiba, M. Izutsu, K. Higuma, J. Ichikawa, T. Lee, V. Filsinger, High-speed dual-parallel Mach-Zehnder modulator using thin lithium niobate substrate, in *Optical Fiber Communication Conference/National Fiber Optic Engineers Conference* (Optical Society of America, Washington, 2008), p. JThA34

24. P.T. Dat, Y. Yamaguchi, K. Inagaki, M. Motoya, S. Oikawa, J. Ichikawa, A. Kanno, N. Yamamoto, T. Kawanishi, Transparent fiber-radio-fiber bridge at 101 GHz using optical modulator and direct photonic down-conversion, in *Optical Fiber Communication Conference* (Optical Society of America, Washington, 2021), p. F3C.4

25. M. Garcia-Granda, H. Hu, J. Rodriguez-Garcia, W. Sohler, Design and fabrica-
    tion of novel ridge guide modulators in lithium niobate. J. Lightwave Technol.
    27(24), 5690–5697 (2009)

26. A. Rao, S. Fathpour, Heterogeneous thin-film lithium niobate integrated photon-
    ics for electrooptics and nonlinear optics. IEEE J. Sel. Topics Quant. Electron.
    24(6), 1–12 (2018)

27. J. Wang, S. Xu, J. Chen, W. Zou, A heterogeneous silicon on lithium niobate
    modulator for ultra-compact and high-performance photonic integrated circuits.
    IEEE Photon. J. **13**(1), 1–12 (2021)

28. P.O. Weigel, F. Valdez, J. Zhao, H. Li, S. Mookherjea, Design of high-bandwidth,
    low-voltage and low-loss hybrid lithium niobate electro-optic modulators. J.
    Phys. Photon. **3**(1), 012001 (2020)

29. Y. Ogiso, J. Ozaki, Y. Ueda, H. Wakita, M. Nagatani, H. Yamazaki, M. Naka-
    mura, T. Kobayashi, S. Kanazawa, Y. Hashizume, H. Tanobe, N. Nunoya, M. Ida,
    Y. Miyamoto, M. Ishikawa, 80-GHz bandwidth and 1.5-V V_π InP-based IQ mod-
    ulator. J. Lightwave Technol. **38**(2), 249–255 (2020)

30. T. Kawanishi, T. Sakamoto, M. Izutsu, K. Higuma, T. Fujita, S. Mori, S. Oikawa,
    J. Ichikawa, 40 Gbit/s versatile LiNbO$_3$ lightwave modulator, in *31st European
    Conference and Exhibition on Optical Communication (ECOC)* (2005)

31. M. Daikoku, I. Morita, H. Taga, H. Tanaka, T. Kawanishi, T. Sakamoto,
    T. Miyazaki, T. Fujita, 100-Gb/s DQPSK transmission experiment without
    OTDM for 100G ethernet transport. J. Lightwave Technol. **25**(1), 139–145
    (2007)

32. P.J. Winzer, G. Raybon, H. Song, A. Adamiecki, S. Corteselli, A.H. Gnauck,
    D.A. Fishman, C.R. Doerr, S. Chandrasekhar, L.L. Buhl, T.J. Xia, G. Wellbrock,
    W. Lee, B. Basch, T. Kawanishi, K. Higuma, Y. Painchaud, 100-gb/s dqpsk
    transmission: from laboratory experiments to field trials. J. Lightwave Technol.
    26(20), 3388–3402 (2008)

33. K. Sato, Y. Kondo, M. Nakao, M. Fukuda, 1.55-μm narrow-linewidth and high-
    power distributed feedback lasers for coherent transmission systems. J. Light-
    wave Technol. **7**(10), 1515–1519 (1989)

34. T. Kunii, Y. Matsui, H. Horikawa, T. Kamijoh, T. Nonaka, Narrow linewidth
    (85 kHz) operation in long cavity 1.55-μm-MQW DBR laser. Electron. Lett.
    27(9), 691–692 (1991)

35. T. Kataoka, K. Hagimoto, Novel automatic bias voltage control for travelling-
    wave electrode optical modulators. Electron. Lett. **27**, 943–945 (1991)

36. T. Pfau, S. Hoffmann, O. Adamczyk, R. Peveling, V. Herath, M. Porrmann,
    R. Noé, Coherent optical communication: towards realtime systems at 40 Gbit/s
    and beyond. Opt. Express **16**(2), 866–872 (2008)

37. C.R.S. Fludger, T. Duthel, D. van den Borne, C. Schulien, E. Schmidt, T. Wuth,
    J. Geyer, E. De Man, G. Khoe, H. de Waardt, Coherent equalization and
    POLMUX-RZ-DQPSK for robust 100-GE transmission. J. Lightwave Technol.
    26(1), 64–72 (2008)

38. A.J. Ward, D.J. Robbins, G. Busico, E. Barton, L. Ponnampalam, J.P. Duck, N.D.
    Whitbread, P.J. Williams, D.C.J. Reid, A.C. Carter, M.J. Wale, Widely tunable

DS-DBR laser with monolithically integrated SOA: design and performance. IEEE J. Sel. Topics Quant. Electron. **11**(1), 149–156 (2005)

39. E. Ip, A.P.T. Lau, D.J.F. Barros, J.M. Kahn, Coherent detection in optical fiber systems. Opt. Express **16**(2), 753–791 (2008)

Chapter 5
Digital Modulation Formats

This chapter describes configurations for optical modulators dedicated to various digital modulation formats. To achieve highly stable optical modulation, deviations from the ideal state, such as optical phase fluctuation and noise effects, must be taken into account, where imperfection of electric signals or imbalance in modulator structures would have impact on transmission performance. Firstly, this chapter describes how modulators with electro-optical effects generate binary modulation format signals which are commonly used for digital communication systems. Simple binary modulation format signal can be generated by simple intensity modulation, where receiver would be insensitive to phase fluctuation. Secondly, multilevel modulation with vector modulation is reviewed, where digital signals with many symbols are generated by electric or optical circuits. For high-speed operation, it would be rather difficult to generate multilevel signals in electric circuits.

5.1 Binary Modulation Formats

Binary modulation uses two types of symbols to transmit one bit described by "0" or "1", which corresponds to one digit in binary notation. Binary modulation formats have been widely used in various transmission systems, and at the same time become elemental technologies for realizing the multilevel modulation formats. Intensity modulation has been widely used because of the simple configuration of modulation and demodulation, but the use of phase modulation has recently been spreading due to the high transmission performance.

5.1.1 Intensity Modulation

The simplest modulation scheme realizes signal transmission with optical signal intensity. In particular, it is called on-off keying (OOK), when it is set to the off-state

© Springer Nature Switzerland AG 2022
T. Kawanishi, *Electro-optic Modulation for Photonic Networks*, Textbooks in
Telecommunication Engineering, https://doi.org/10.1007/978-3-030-86720-1_5

for "0" and the on-state for "1". Figure 5.1 shows the OOK signal where the symbols consist of amplitude "0" and amplitude "1".

If the amplitude of the output light is unity, the distance between the two symbols is also equal to unity. At the receiver side, the locations of the symbols would be shifted, and the symbols should have some divergence due to noise generated by amplifiers in the transmission systems. Bit errors can be suppressed, when the separation between symbols is large. The optical phase of the on-state is zero, that is, it is on a positive real axis, but in an actual optical communication system, the optical phase varies greatly as described in Sect. 4.5.

Figure 5.2 shows a constellation taking the phase fluctuation into account. At the receiver side, the light intensity, i.e., the distance from the origin on the phasor diagram is detected, to determine the on-state in which the intensity is equal to or greater than a predetermined value. Signal transmission is possible without being affected by the optical phase fluctuation, in principle. The optical intensity does not change even if the phase change occurs by the chirp effect at the transmitter side, but the phase change is converted to the intensity change by the dispersion effect during the fiber transmission, which causes the degradation of the received waveform.

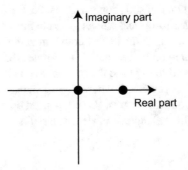

Fig. 5.1 Constellation of OOK signal

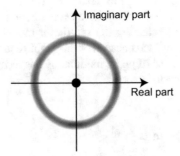

Fig. 5.2 Constellation of OOK signal with optical phase fluctuation

Figures 5.3 and 5.4 show the transition states of the optical signals during on-off switching. When the chirp is almost zero, the length of the trajectory is the minimum as shown in Fig. 5.3. On the other hand, when the chirp is large, the path length on the phasor diagram is very long, where the trajectory of the optical signal transition state has the rotation component corresponding to phase rotation associated with intensity change. When the modulation speed is the same, the higher the chirp, the higher the rate of change of the optical signal state on the phasor plane. The spectrum width of the optical signal depends on the speed of the state transition, where the faster the speed, the larger the spectrum width. The optical signal generated by the direct modulation has the long trajectory as shown in Fig. 5.4, while the chirp of the MZM is small enough to offer the zero-chirp intensity modulation whose trajectory has the shortest path shown in Fig. 5.3. Thus, we can use the direct modulation for short distance or low baud-rate transmission where the optical fiber dispersion effect is not so large. The low- or zero-chirp amplitude modulation is useful for enhancement of transmission length or spectral efficiency.

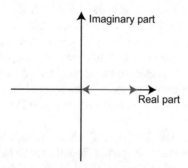

Fig. 5.3 Transition state trajectory of OOK signal without chirp

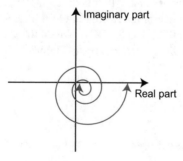

Fig. 5.4 Transition state trajectory of OOK signal with chirp

Here, we consider the OOK signal waveforms generated by actual MZMs. The waveform largely depends on the MZM bias condition. Figure 5.5 (left) shows the relationship between the bias and symbol of the response curve (response of the output intensity to the voltage applied to the modulator). Suppose that the voltage change between two symbols is less than the half-wave voltage $V_{\pi MZM}$, which is the voltage required to change the light intensity from the minimum to the maximum. If the light intensity is half at point C as bias point, the constellation as shown in Fig. 5.5 (right) on the phasor plane is achieved, where points A and B are symbols. This corresponds to the quadrature bias defined in Sect. 4.3.3. The applied voltage has a constant fluctuation due to noise and incompleteness in the electric circuit for generating and boosting the modulation signal. The optical intensity varies according to the noise, so that the symbols on the phasor diagram have fluctuation in the radius direction. Noise accumulated through the optical transmission by the optical amplifiers, etc. causes fluctuation in both the radius direction and rotation direction. For simplicity, the chirp parameter α_0 of the MZM is assumed to be zero. If there is a chirp, there is an optical phase difference between points A and B, and the three points A, B, and C are not aligned on a straight line, but the response curve in Fig. 5.5 (left) does not change. As shown in Fig. 5.2, the OOK signal with the optical phase fluctuation has a donut-shaped constellation, where the chirp does not affect the radius, i.e., the intensity. Figure 5.6 shows the case where the symbol (point A) is set to the minimum point. The symbol positions on the optical intensity curve can be controlled by the bias point C which depends on the applied DC voltage. At the point A, the derivative with respect to the applied voltage is zero, so that the optical intensity variance due to the fluctuation in the applied voltage can be suppressed, where the variance of the point A is much smaller than that of the point B as shown in Fig. 5.6 (right).

If the quadrature bias condition (point C where the light intensity is half) and the voltage difference between the two symbols is equal to the half-wave voltage $V_{\pi MZM}$, the optical intensity is maximal and minimal at the points A and B, respectively, as shown in Fig. 5.7. In addition, the change in the optical intensity with respect to the applied voltage at both points is minimal, and the influence of the voltage fluctuation can be suppressed. This bias condition is suitable for the OOK signal generation because it is compatible with maximizing the distance between symbols and minimizing the influence of the fluctuation of the electric signal. This is one of the major advantages of the MZMs for OOK, where the intensity fluctuation can be largely suppressed at symbols where the optical intensity is maximal and minimal.

5.1.2 Phase Modulation

This section describes binary-phase-shift-keying (BPSK) that uses two different phase states. Although it is often referred to simply as phase-shift-keying (PSK), various modulation schemes have been developed in recent years. In order to avoid confusion, it is desirable to refer to PSK as a generic digital phase modulation scheme, i.e., binary one as BPSK and four-valued phase modulation quadrature-

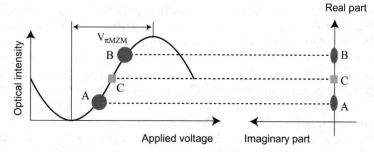

Fig. 5.5 OOK signal generation by an MZM. (left) two symbols and the bias point on the MZM response curve. (right) OOK signal constellation where the vertical axis shows the real component

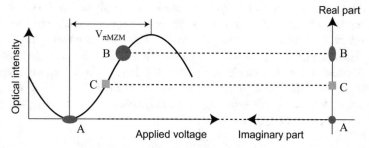

Fig. 5.6 OOK signal generation by an MZM, where one of the symbols is on the optical intensity minimum. (left) two symbols and the bias point on the MZM response curve. (right) OOK signal constellation where the vertical axis shows the real component

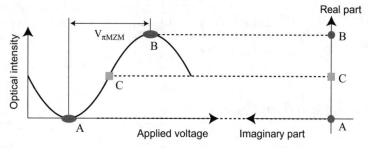

Fig. 5.7 OOK signal generation by an MZM, where the bias point and the amplitude of the modulating signal are optimized. (left) two symbols and the bias point on the MZM response curve. (right) OOK signal constellation where the vertical axis shows the real component

phase-shift-keying (QPSK), and to refer to phase modulation using more symbols as n-PSK (where n is the number of symbols).

Figure 5.8 shows the constellation of a BPSK signal. The amplitude equals unity. The two states whose phases are 0 and 180 degrees are used as symbols. The distance between symbols is twice that of OOK. Other values, such as 90 degrees of phase change, are also possible in principle. Usually 360 degrees are equally partitioned for symbol placement, to maximize distances between symbols is necessary to improve transmission performance. The phase difference is 180 degrees for BPSK, while that of QPSK is 90 degrees. 180 degree phase change corresponds to inverting the sign of the amplitude, and the two symbols are both on the real axis, as shown in Fig. 5.8 This means that BPSK with a 180-degree phase difference is equivalent to amplitude modulation (amplitude-shift-keying: ASK) with two amplitude symbols ±1. Thus, BPSK signals can be generated by amplitude modulators. The constellation of the BPSK signals is shown in Fig. 5.9 taking the optical phase fluctuation into account. If modulation speed is faster than fluctuating speed, the phase relationship between the preceding and following symbols is kept at 0 degrees or 180 degrees. Thus, the effects of phase rotation can be suppressed by detecting the phase difference between preceding and following symbols. In differential-phase-shift-keying (DPSK), the phase difference is used as the symbol instead of the phase itself. For example, the 0-degree phase difference is used as the symbol for "0", while the 180-degree phase difference is associated with "1'. The DPSK format with two symbols which called differential-binary-phase-shift-keying (DBPSK) and that with four symbols which is called differential-quadrature-phase-shift-keying (DQPSK) are often used in actual transmission systems [1, 2], where the method of detecting the phase difference between the symbols having the time difference at the receiver side by the interference is called delay detection.

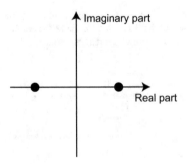

Fig. 5.8 Constellation of a BPSK signal

Here, we discuss the generation of BPSK signals using zero-chirp MZMs as amplitude modulators. The optical output signals of the MZMs move on the real axis, when the phase fluctuation can be neglected. Figure 5.10 (left) shows the symbols and bias points on the amplitude response curve with respect to the applied voltage (the relationship between the applied voltage and the output optical amplitude). The bias is set to the null bias where the light intensity is minimal (see Sect. 4.3.3). The MZM inverts the polarity of the optical amplitude to provide two symbols with 180-degree phase difference, where the bias point is at the optical intensity minimum. Such bias condition is called the null-bias. While the modulating signal amplitude is equal to $V_{\pi MZM}$ for OOK, BPSK requires a voltage change of $2V_{\pi MZM}$ to maximize the inter-symbol distance, as shown in Figs. 5.7 and 5.10.

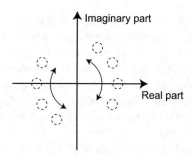

Fig. 5.9 Constellation of a BPSK signal with phase fluctuation

BPSK signals can also be generated by optical phase modulators. Figure 5.11 (left) shows the relationship between two symbols and bias on the response curve for optical phase. Since the fluctuation of the applied voltage leads to the variation of the phase of the output light, the symbol in the angular direction on the phasor diagram is expanded as shown in Fig. 5.11 (right). On the other hand, when the BPSK signal is generated by the MZM, the absolute value of the optical signal becomes the largest (i.e., the optical signal intensity becomes the largest) at the symbols (point A and B), where the variation of the amplitude due to the fluctuation of the applied voltage can be suppressed, as with OOK shown in Fig. 5.7.

Figure 5.12 shows transient states between the symbols of BPSK signals as a trajectory. In the case of the MZM (left), it moves on a straight line connecting the symbols with the shortest distance. In the case of the phase modulator (right), the path of the transient state becomes an arc, and the path length becomes large. Therefore, the MZM can reduce the speed of the change of the optical signal state on the phasor diagram and suppress the spectrum broadening. Moreover, in the high-speed transmission system, the state does not stay long at the symbols, and the transient state would take up the majority when viewed on the time axis.

Figure 5.13 shows a time domain BPSK signal waveform generated by the MZM. During the transient, the phase is constant at 0 or 180 degrees, and varies rapidly when the state passes through the origin. On the other hand, the amplitude changes greatly in the transient state. The intensity is minimal in the middle of the transient state (zero for ideal modulators). This has an effect of suppressing transient optical energy that does not contribute to signal transmission, and is a property suitable for improving transmission characteristics. Since the trajectory of the BPSK signal generated by the MZM is symmetric with respect to the origin, the average of the states including the transient is on the origin. This means that the optical signal does not have any DC components.

On the other hand, when the BPSK signal is generated by the phase modulator shown in Fig. 5.14, the intensity is constant and the phase varies continuously from 0 degrees to 180 degrees. The state moves only the first and second quadrants in the phasor plane. Since the path is symmetrical to the imaginary axis, but the imaginary component is skewed to the positive range, the temporal average is a constant value on the imaginary axis and the positive range. This means that the optical signal has a DC component in the phasor representation, that corresponds the optical carrier component. In principle, it is possible to use signal processing on the electric signal to be applied to the electrode in advance so as to make the trajectory pass through the third and fourth quadrants, but it is difficult to apply the signal to high-speed modulation because twice the electric signal amplitude and complicated digital processing are required. In the phase modulation, it is desirable to suppress the carrier components that do not directly contribute to information transmission at the transmitter side. However, the transient part is not negligible in the fast modulation, so that the carrier component with 90 degree phase deviation remains in the optical output in the BPSK signals by the phase modulator.

Figure 5.15 compares the BPSK signal spectra generated by the phase modulator and the MZM. In the MZM, if the bias point is set appropriately, the trajectory of the optical signal state becomes symmetric with respect to the origin, and the carrier can be easily suppressed, which is advantageous in terms of transmission characteristics. As noted above, both MZ and phase modulators can generate BPSK signals, but MZMs can suppress the following three elements that affects transmission performance, compared to phase modulators: (1) the effects of fluctuations in electrical signals, (2) energy with transient parts, and (3) energy of carrier components. These are summarized in Table 5.1.

Since the MZM shifts symbols at the shortest in the phase diagram, unnecessary spectrum broadening can be suppressed. The optical loss in the MZM is generally larger than that of the phase modulator because of the loss at the branch and the component that is theoretically emitted at the junction. However, because the symbols of the BPSK signals are at the maximum transmission of the MZM, the negative impact of the optical loss is relatively small. In addition, MZMs are more complex than phase modulators and require bias control, but they are commonly used in actual PSK-based transmission systems to enhance the transmission performance. As described in Sect. 5.2, signal generation techniques are being developed that combines two or more BPSK signals by the MZMs, for high-speed transmission systems with multilevel modulation formats.

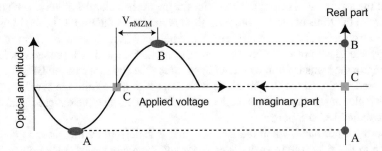

Fig. 5.10 BPSK signal generation by the MZM. (left) two symbols and the bias point on the MZM response curve. (right) BPSK signal constellation where the vertical axis shows the real component

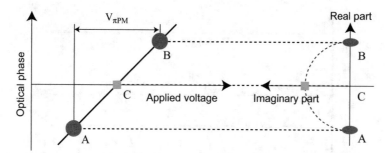

Fig. 5.11 BPSK signal generation by the phase modulator. (left) two symbols and the bias point on the phase modulator response curve. (right) BPSK signal constellation where the vertical axis shows the real component

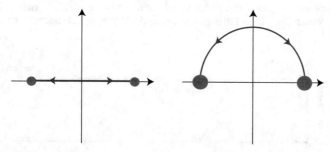

Fig. 5.12 Symbols and transition of BPSK signals. (left) BPSK signal generation by the MZM, where the transient states are also only on the real axis as well as symbols. (right) BPSK signal generation by the phase modulator, where the transient states are on the circle whose intensity is constant

Table 5.1 Comparison between MZM and phase modulator

	MZM	Phase modulator
Suppression of amplitude fluctuation	Yes	No
Suppression of transient components	Yes	No
Suppression of optical carrier	Excellent	Fair
Shortest path between symbols	Yes	No
Bias control free	No	Yes

5.1.3 Frequency Modulation

Optical frequency can be controlled by using combination of phase modulation signals generated by the EO effect. This section describes the binary frequency-shift-keying (FSK), which is a frequency modulation format with two symbols associated with two particular optical frequencies.

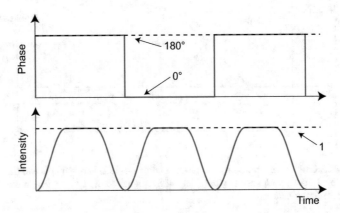

Fig. 5.13 Time domain profile of a BPSK signal generated by the MZM. (upper) optical phase, which changes from 0 degree to 180 degree at the origin. (lower) optical intensity, which as a dip at the origin

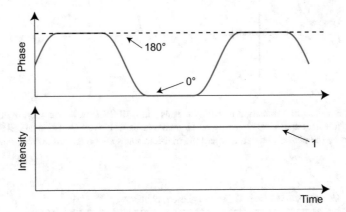

Fig. 5.14 Time domain profile of a BPSK signal generated by the phase modulator. (upper) optical phase, which changes from 0 degree to 180 degree continuously. (lower) optical intensity, which is constant on the path as a whole

The phasor diagram is based on the optical frequency of the input lightwave, i.e., the carrier frequency. Thus, when the frequency of the signal is equal to the carrier frequency, the signal can be described as a point on the phasor plane. Here, we consider an FSK signal with two symbols of two frequencies, $(\omega_0 + \omega')/2\pi$ and $(\omega_0 - \omega')/2\pi$, which can be described by $e^{i(\omega_0+\omega')t}$ and $e^{i(\omega_0-\omega')t}$. By using the phasor plane based on the optical carrier whose sinusoidal oscillation term is $e^{i\omega_0 t}$, the two symbols can be expressed by

$$e^{\pm i\omega' t} = \cos \omega' t \pm i \sin \omega' t. \tag{5.1}$$

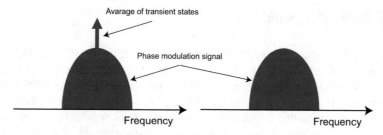

Fig. 5.15 Spectra of BPSK signals, (left) generated by the PM, and (right) generated by the MZM

As shown in Fig. 5.16, the symbol with the higher frequency $(\omega_0 + \omega')/2\pi$ corresponds a counterclockwise circular trajectory on the phasor plane, while the symbol with the lower frequency $(\omega_0 - \omega')/2\pi$ corresponds a clock wise circular trajectory. At the receiver side, a simple direct detection by a photodetector which converts optical intensity change into electric current cannot be used for such FSK signals, since the optical intensity is constant. Frequency discrimination or coherent demodulation should be used for FSK signal demodulation. Optical filters or delay interferometers can offer frequency discrimination with simple configuration, however, temperature control or mechanical vibration mitigation would be needed for stable operation.

Equation (5.1) shows that an FSK signal can be generated by amplitude modulation for the real component with $\cos \omega' t$ and the imaginary component with $\sin \omega' t$, i.e., the vector modulation by $\cos \omega' t$ and $\sin \omega' t$ [3, 4]. As described in Sect. 7.2.2, the principle of the FSK signal generation can be derived from sideband generation at each phase modulator. The optical frequency control by direct laser modulation can be used for FSK [5, 6]. However, there are some problems such as parasitic optical intensity change in frequency control [7].

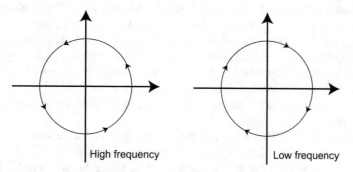

Fig. 5.16 FSK signal on the phasor plane

5.2 Multilevel Modulation

Multilevel modulation can offer transmission more than one bit of information in a single modulation time slot, by using three or more symbols, where 2^N symbols are required for N-bit transmission in general. In commercial systems, QPSK is often used to increase the spectral efficiency and transmission capacity simultaneously.

Multilevel modulation formats using only amplitude or intensity are possible, however, it is common to use vector modulation because the distance between symbols can be increased in the phasor plane by using phase control. This chapter describes multilevel modulation schemes using quadrature amplitude modulation (QAM) by parallel MZMs shown in Sect. 4.4.1.

5.2.1 Quadrature Phase Shift Keying

Figure 5.17 shows the constellation of the PSK signal with four symbols which have 90-degree separation in phase, with constant optical intensity. One modulation time slot of the PSK signal transmits 2 bits. When the optical intensity of the signal is unity, the separation of the symbols in the phasor plane is $\sqrt{2}$ which is larger than in OOK (the separation is unity in OOK). This phase modulation is called 4-PSK or QPSK because it uses four symbols with difference optical phases.

However, in the QAM expressed by (4.74), the signal of the same constellation is produced if $X, Y = \pm 1$, where the optical intensity is normalized where the intensity is unity. This QAM signal has four symbols, so that it can be called 4-QAM. This means that QPSK and 4-QAM refer to the same modulation scheme.

As shown in Fig. 5.18, a phase modulator can generate a QPSK signal from four level amplitude modulation electric signal, as with BPSK signal generation by a phase modulator where a binary electric signal is fed to the modulator (see Fig. 5.11). On the other hand, a QAM signal with $(X, Y) = (\pm 1, \pm 1)$ can be generated from two BPSK signals (2-ASK signals with ± 1) by the MZMs, with 90-degree phase difference between the two optical signals on the real and imaginary axes, as shown in Fig. 5.19. These two BPSK signals can be combined together by an optical signal combiner (Y-junction) integrated in the dual-parallel MZM. The combiner can offer vector summation, so that a QPSK (4-QAM) signal can be obtained as an optical output of the modulator.

Although the BPSK signal can be obtained by a phase modulator, as described above, a dual-parallel MZM with two MZMs integrated in parallel as shown in Fig. 5.19 is often used for generating the QPSK signal because the MZMs has advantages such as suppression of fluctuation of the electric signal as described in Sect. 5.1.2 [1, 2, 8]. QPSK has a balanced property that the distances between symbols are greater than OOK and that two bits of information can be transmitted in a single modulation time slot.

Figure 5.20 shows the transitions between symbols and symbols of QPSK signals as trajectories. In the case of a dual-parallel MZM (left), the paths are on straight lines connecting the symbols with the shortest distances. In the case of a phase modulator (right), the paths of the transient states become arcs, and the movement distances become long. In addition, due to the necessity of high-speed digital signal processing and large amplitude voltage, it is difficult to offer QPSK signal paths which distributes uniformly over the circle on the phasor plane as in the case of BPSK, so that the carrier component is left as shown in Fig. 5.15 (left). An electric signal with four different voltage levels is applied to the phase modulator. It is rather difficult to generate multilevel electric signals for high-speed modulation, where high-speed digital-to-analog converters (DACs) and linear amplifiers are required. On the other hand, the process of generating signals by the QAM shown in Fig. 5.19 utilizes two degrees of freedom in the optical signal: a real and imaginary component, where binary amplitude modulation is applied. The optical output, which is synthesized from two binary modulated signals, is a 4-QAM signal. As described above, the 4-QAM signal can be regarded as a QPSK signal with four-valued phases as symbols. However, the signal generation by the dual-parallel MZM does not have any multilevel signals in the electric circuit or within the modulator and is achieved by combining two binary modulation signals.

Thus, it is possible to utilize conventional technologies such as electronic circuits and MZMs developed for binary modulation formats, so it would be a practical technology although the complexity of bias control and the need for integrated dual-parallel MZMs are issues. The waveform control preciseness required for electronic circuits and modulators is comparable to that of BPSK and OOK (Table 5.2).

5.2.2 Synthesis of Multilevel Signals

QPSK (or 4-QAM) signals can be synthesized from two binary modulation signals on the real and imaginary components, where the preciseness for optical waveform control required by the transmitter is basically comparable to that for conventional bi-

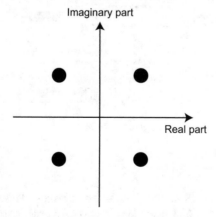

Fig. 5.17 Constellation of QPSK

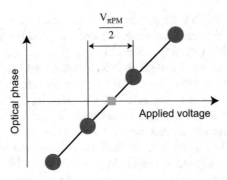

Fig. 5.18 QPSK by phase modulation

Table 5.2 Comparison between dual-parallel MZM and phase modulator

	Dual-parallel MZM	Phase modulator
Suppression of amplitude fluctuation	Yes	No
Suppression of transient components	Good	Poor
Suppression of optical carrier	Excellent	Fair
Shortest path between symbols	Yes	No
Bias control free	No	Yes
Simplicity of device	Fair	Excellent
Multilevel signal free	Yes	No

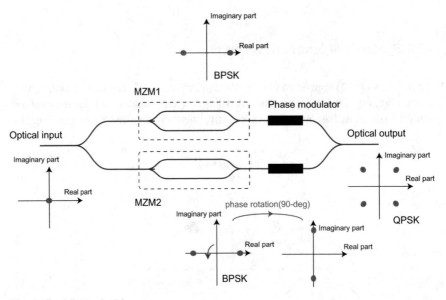

Fig. 5.19 QPSK by QAM

nary modulation formats. However, there are some issues on fabrication of integrated modulators, and on control of many bias voltages. In order to realize more complex

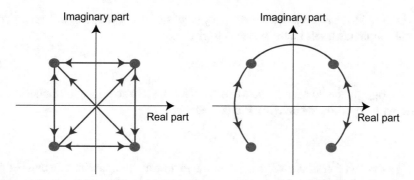

Fig. 5.20 Constellation and transient states of QPSK signals. (left) QPSK signal generation by the dual-parallel MZM, where the transient states are on the shortest lines connecting the symbols. (right) QPSK signal generation by the phase modulator, where the transient states are on the circle whose radius corresponds to the optical signal intensity

modulation schemes, it is necessary to apply multi-valued amplitude modulation to the real and imaginary components.

The most straightforward approach is to apply multilevel modulation to X and Y in the QAM given by (4.74), where a 16-QAM signal can be generated by using symbols of $\pm 1, \pm 1/3$ for X and Y, as shown in Fig. 4.60. In this case, symbols which are not on the intensity maxima should be used, so that the effect of the amplitude fluctuation suppression would be weaker than in QPSK. In addition, multilevel electric signals should be generated in the electric circuits to drive the modulator as with the QPSK using the phase modulator. Thus, some slight imbalance in the modulators, such as the intrinsic chirp would be issues, where precise waveform generation and management are very important in electric and optical components.

As shown in Fig. 5.21, complicated signals can be obtained by superposition of signals with binary modulation signals. For example, a 16-QAM signal can be generated by combining two QPSK signals with 6 dB amplitude difference. Figure 5.22 shows a configuration of 16-QAM signal generation by using a quad-parallel MZM (QPMZM) which consists of two DPMZMs, where the amplitudes of the optical signals are controlled in the integrated optical circuits [9, 10]. Each dual-parallel MZM generates a QPSK signal. Since each QPSK signal consists of two BPSK signals, the 16-QAM signal is synthesized from four BPSK signals combined together in the QPMZM. The required amplitude differences between two QPSK signals can be maintained by controlling the power splitting ratio at the Y-junctions or loss of the optical circuits. The amplitude is also can be controlled by the positions of the BPSK symbols on the MZM response curves. In the former case, there is a problem that the optical circuit becomes complicated, but electric signal fluctuation can be largely suppressed by using a maximum point of the MZM intensity response as a symbol. In the latter case, we can use simple optical devices, but the above-mentioned fluctuation suppression effect cannot be achieved. In either case, we can use conventional low-cost devices dedicated to binary modulation formats, to drive the modulators.

QAM with many levels can be offered by large-scale integrated MZIs. The number of the amplitude levels on the real axis is given by

$$N_{\mathrm{R}} = 2^{N_{\mathrm{MZr}}}, \tag{5.2}$$

where N_{MZr} is the number of the MZIs used for the real component. Similarly, the number of levels on the imaginary axis is equal to

$$N_{\mathrm{I}} = 2^{N_{\mathrm{MZi}}}. \tag{5.3}$$

N_{MZi} is the number of the MZIs used for the imaginary component. The number of the symbols in QAM can be given by a product of the numbers of levels on the real and imaginary axes, as follows:

$$\begin{aligned} N_{\mathrm{QAM}} &= 2^{N_{\mathrm{MZr}}} \times 2^{N_{\mathrm{MZi}}} \\ &= 2^{N_{\mathrm{MZr}}+N_{\mathrm{MZi}}}. \end{aligned} \tag{5.4}$$

By using the total number of the MZI embedded in the parallel MZM, which is defined by

$$N_{\mathrm{MZ}} = N_{\mathrm{MZr}} + N_{\mathrm{MZi}}, \tag{5.5}$$

the number of symbols in QAM can be expressed by

$$N_{\mathrm{QAM}} = 2^{N_{\mathrm{MZ}}}. \tag{5.6}$$

In common QAM modulation schemes, the numbers of amplitude levels in the real and imaginary components are equal to each other, where

$$N_{\mathrm{MZr}} = N_{\mathrm{MZi}} = N_{\mathrm{MZ}}/2. \tag{5.7}$$

In BPSK by an MZM, $N_{\mathrm{MZr}} = 1$, while $N_{\mathrm{MZi}} = 0$. Thus, the total number of symbols is equal to 2. In QPSK by a DPMZM, $N_{\mathrm{MZr}} = N_{\mathrm{MZi}} = 1$, i.e., $N_{\mathrm{MZ}} = 2$, so that the total number of the symbols is equal to 4. Here, for example, we consider an octa-parallel MZM as shown in Fig. 5.23, where $N_{\mathrm{MZr}} = N_{\mathrm{MZi}} = 4$, i. e. $N_{\mathrm{MZ}} = 8$. As described in (5.6), a 256-QAM signal shown in Fig. 5.24, can be generated from 8-lane binary bit streams [11]. However, there are problems such as large device size and large optical loss. In addition, it is rather difficult to deal with complicated bias control. The octa-parallel MZM has 16 integrated phase modulators and requires 15-dimensional bias control. Recently, technologies for synthesizing high-speed multilevel signals with electric circuits have also been developed, and applied to complex signal generation with a DPMZM [12]. It is also possible to optimize the total performance of QAM by combining the synthesis in the optical circuit with the multilevel signal generation in the electric circuit.

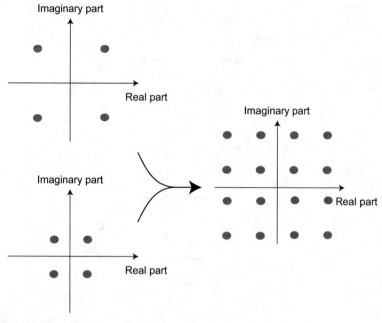

Fig. 5.21 16-QAM signal generation from two QPSK signals

Problems

5.1 Consider 16-level phase-shift-keying by an optical phase modulator, whose half-wave voltage is 8.0 V. What is the voltage step in the modulating signal?

5.2 Consider QAM signal generation by a parallel MZM with parallel bit streams. How many MZIs are required for 1024-level QAM?

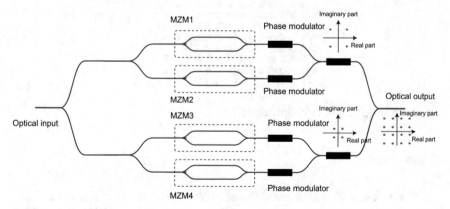

Fig. 5.22 16-QAM signal generation by quad-parallel MZM

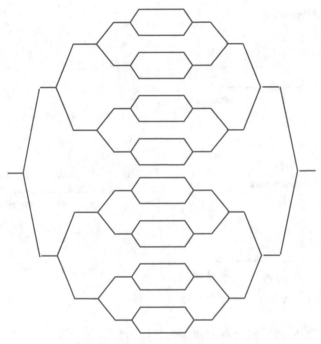

Fig. 5.23 Schematic of octa-parallel MZM

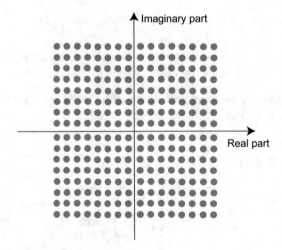

Fig. 5.24 Constellation of a 256-QAM signal

References

1. T. Kawanishi, T. Sakamoto, M. Izutsu, High-speed control of lightwave amplitude, phase, and frequency by use of electrooptic effect. IEEE J. Select. Top.

Quantum Electron. **13**(1), 79–91 (2007)

2. M. Daikoku, I. Morita, H. Taga, H. Tanaka, T. Kawanishi, T. Sakamoto, T. Miyazaki, T. Fujita, 100-Gb/s DQPSK transmission experiment without OTDM for 100G ethernet transport. J. Lightwave Technol. **25**(1), 139–145 (2007)

3. T. Kawanishi, K. Higuma, T. Fujita, J. Ichikawa, T. Sakamoto, S. Shinada, M. Izutsu, $LiNbO_3$ high-speed optical FSK modulator. Electron. Lett. **40**(11), 691–692 (2004)

4. M. Izutsu, S. Shikamura, T. Sueta, Integrated optical SSB modulator/frequency shifter. IEEE J. Quantum Electron. **17**(11), 2225–2227 (1981)

5. D.A. Fishman, Performance of single-electrode 1.5-μm DFB lasers in nonco-herent FSK transmission. J. Lightwave Technol. **9**(7), 924–930 (1991)

6. H. Tsushima, S. Sasaki, K. Kuboki, S. Kitajima, R. Takeyari, M. Okai, 1.244-Gb/s 32-channel transmission using a shelf-mounted continuous-phase FSK optical heterodyne system. J. Lightwave Technol. **10**(7), 947–956 (1992)

7. A.E. Willner, M. Kuznetsov, I.P. Kaminow, J. Stone, L.W. Stulz, C.A. Bur-rus, FM and FSK response of tunable two-electrode DFB lasers and their performance with noncoherent detection. IEEE Photon. Technol. Lett. **1**(12), 412–415 (1989)

8. T. Kawanishi, T. Sakamoto, M. Izutsu, K. Higuma, T. Fujita, S. Mori, S. Oikawa, J. Ichikawa, 40 Gbit/s versatile $LiNbO_3$ lightwave modulator, in *31st European Conference and Exhibition on Optical Communicatio (ECOC)*, 2005

9. T. Sakamoto, A. Chiba, T. Kawanishi, 50-Gb/s 16 QAM by a quad-parallel Mach-Zehnder modulator, in *33rd European Conference and Exhibition of Optical Communication - Post-Deadline Papers (published 2008)* (2007), pp. 1–2

10. T. Kawanishi, T. Sakamoto, A. Chiba, Integrated lithium niobate Mach-Zehnder interferometers for advanced modulation formats. IEICE Trans. Electron. **E92.C**(7), 915–921 (2009)

11. T. Kawanishi, Parallel Mach-Zehnder modulators for quadrature amplitude modulation. IEICE Electron. Expr. **8**(20), 1678–1688 (2011)

12. M. Yoshida, H. Goto, K. Kasai, M. Nakazawa, 64 and 128 coherent QAM optical transmission over 150 km using frequency-stabilized laser and heterodyne PLL detection. Opt. Express **16**(2), 829–840 (2008)

Chapter 6
Sideband Generation by Optical Modulation

This chapter describes the operating principles of the various modulators based on optical phase modulation, by using mathematical expressions with the Bessel function, which is one of the most well-known special functions for solving the wave equations of cylinder waves. The Bessel function is also very useful to describe phase modulation signals. Assuming that the wave is represented by a trigonometric function, phase modulation is equivalent to changing the variable of the trigonometric function by another trigonometric function. When learning trigonometric functions with spreadsheets or functional calculators, did you try to apply the sine operation twice on a number? The Bessel functions can provide concise expressions for such mathematical operations in a prospective way. This chapter introduces the basic properties of the Bessel function and discusses the operation principles of the various modulators and their characteristics.

6.1 Mathematical Expression for Phase Modulation by Bessel Function

This section provides mathematical expressions of phase modulation with sinusoidal signals. Optical phase modulation, which corresponds to changing a variable of a sine, cosine, or complex-valued exponential function with a sine or cosine waveform, for example, $\sin\{\sin(x)\}$, can be described by a mathematical expansion with the Bessel function. Each component in the expansion corresponds to an optical frequency whose frequency difference from the optical input is an integer multiple of the modulation signal frequency. Such components are called sidebands. In addition to the mathematical expression with the Bessel function for phase modulation signals, this chapter describes the properties of the Bessel functions that are useful for understanding optical modulation. The addition theorem of the Bessel functions is important to understand the additivity of phase modulation.

© Springer Nature Switzerland AG 2022
T. Kawanishi, *Electro-optic Modulation for Photonic Networks*, Textbooks in
Telecommunication Engineering, https://doi.org/10.1007/978-3-030-86720-1_6

6.1.1 Basics of Phase Modulation Expressed by Bessel Function

For preparation of detailed discussion on various modulators consisting of two or more optical phase modulators, the optical output of the i-th optical phase modulator (R_i) is expressed by

$$R_i = K_i e^{i\omega_0 t + i v_i(t)},\tag{6.1}$$

as with Eq. (4.20), where K_i denotes the product of the optical input amplitude E_0, multiplied by the loss of the i-th optical phase modulator K_{Li} [1]. Assuming that the modulating signal v_i is a sinusoidal wave signal expressed by

$$v_i(t) = A_i \sin(\omega_m t + \phi_i) + B_i,\tag{6.2}$$

the output optical signal can be given by

$$R_i = K_i e^{i\omega_0 t} e^{i A_i \sin(\omega_m t + \phi_i)} e^{i B_i}.\tag{6.3}$$

B_i is the DC component of the modulating signal. The initial phase of the output signal can be shifted by changing B_i. The interference between the output signals in the modulators, such as the MZM, consisting of more than one phase modulators is affected, where the optical output signal is synthesized from more than one phase modulated signals. In general, B_i is called the bias of the modulator. Since the output optical intensity depends only on the phase difference for MZMs, the bias for the MZMs consisting of phase modulators i and j is often referred to as $B_i - B_j$.

Similarly, in the modulators consisting of more than two phase modulators, the initial phases of the optical signals before combing the signals for output signal generation can be given by the term of $e^{i B_i}$, so that the output signal intensity depends only on each phase difference.

$A_i \sin(\omega_m t + \phi_i)$ describes the sinusoidal modulating signal with rapid oscillation whose frequency is from several MHz to several 10 GHz. The signal components that vary rapidly are referred to as radio-frequency (RF) signals as appropriate.

Nonlinear effects, such as sideband generation due to phase modulation, can be expressed by

$$r_i = e^{i A_i \sin(\omega_m t + \phi_i)}.\tag{6.4}$$

By using real functions, it is given by

$$\mathrm{Re}[r_i] = \cos\left[A_i \sin(\omega_m t + \phi_i)\right].\tag{6.5}$$

A composite function of $\cos x$ and $\sin x$ can be expressed by an expansion with the Bessel function (the first kind Bessel function) $J_n(z)$ as follows:

$$\cos(z \sin \theta) = \sum_{k=-\infty}^{\infty} J_k(z) \cos k\theta.\tag{6.6}$$

By using this expansion, the phase modulated signal can be given by

$$\mathrm{Re}[r_i] = \sum_{n=-\infty}^{\infty} J_n(A_i) \cos(n\omega_m t + n\phi_i). \qquad (6.7)$$

The n-th component shows that the amplitude of the sideband component shifted from the optical carrier by the frequency of $n\omega_m/2\pi$ is given by the Bessel function $J_n(A_i)$. Details on the expansion with Bessel function will be discussed in Sect. 6.1.4.

6.1.2 Basic Properties of Bessel Functions

The n-th order the first kind Bessel function $J_n(z)$ is defined by

$$J_n(z) = \sum_{m=0}^{\infty} \frac{(-1)^m}{m!(n+m)!} \left(\frac{z}{2}\right)^{n+2m}, \qquad (6.8)$$

which is a solution of the Bessel differential equation:

$$\frac{\mathrm{d}^2 u}{\mathrm{d}z^2} + \frac{1}{z}\frac{\mathrm{d}u}{\mathrm{d}z} + \left(1 - \frac{n^2}{z^2}\right)u = 0. \qquad (6.9)$$

The first kind Bessel functions are often simply called the Bessel functions (in a narrow sense) [2, 3]. n is 0 or a positive integer. Although it is possible to define a Bessel function that extends the order n to a real number, the phase modulated signal can be perfectly expressed only by the Bessel functions with the integer orders. For negative integer n, the Bessel function can be defined by

$$J_{-n}(z) \equiv (-1)^n J_n(z), \qquad (6.10)$$

so that the phase modulated signal can be expressed by using the Bessel functions with 0 or positive integer orders.

Figure 6.1 shows the curves of the first kind Bessel functions $J_n(z)$ of $n = 0, 1, 2, 3, 4$, where z is from 0 to 10 on the real axis. In this chapter, the modulation signal to be applied to the modulator is represented as a real function, so the z is assumed to be a real number. By using (6.8), the Bessel functions can be approximately expressed by

$$J_n(z) \simeq \frac{1}{n!} \left(\frac{z}{2}\right)^n, \qquad (6.11)$$

when z is close to 0. Thus, $J_n(z)$ of $n = 0, 1, 2, 3, 4$ can be given by

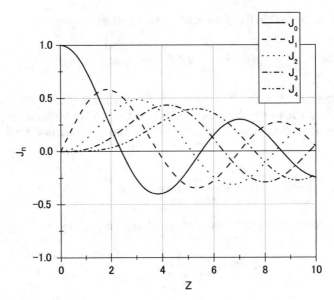

Fig. 6.1 The first kind Bessel functions

$$J_0(z) \simeq 1 \tag{6.12}$$

$$J_1(z) \simeq \frac{z}{2} \tag{6.13}$$

$$J_2(z) \simeq \frac{z^2}{8} \tag{6.14}$$

$$J_3(z) \simeq \frac{z^3}{48} \tag{6.15}$$

$$J_4(z) \simeq \frac{z^4}{384}. \tag{6.16}$$

$J_0(z)$ is an even function that has a maximum of 1 at $z = 0$. $J_n(z)$ $(n \neq 0)$ are odd or even functions, which are equal to 0 at $z = 0$ and are proportional to z^n when $z \sim 0$.

For J_0, the approximation by the first term gives just a constant value, so that the approximation using the first two terms $(m = 0, 1)$,

$$J_0(z) = 1 - \frac{z^2}{4}, \tag{6.17}$$

would be useful. Figs. 6.2, 6.3, and 6.4 show comparisons between exact solutions and approximate expressions, for J_0, J_1, and J_2, respectively. The error is defined as the difference between the approximate value and the exact value, normalized with the exact value.

In the well-balanced push–pull operation using an MZM with an x-cut LN substrate, the required optical phase shift at each phase modulator is $\pi/2$ as a peak-to-peak value, where the bias is set for the optimal OOK signal generation with the largest optical output at the on-state. In this case, the phase shift described by A_i in (6.7) should be in the range from 0 to 0.8 ~ $\pi/4$, approximately, where the errors for J_0, J_1, and J_2 shown in Figs. 6.2, 6.3, and 6.4 are smaller than 0.7%, 8.4%, and 6.3%, respectively. Thus, these approximations can be used for rough estimation of the values of the Bessel function when z is in the range from 0 to 0.8 ~ $\pi/4$, where z is smaller than 0.8 ~ $\pi/4$ in most of the modulator applications. When a larger modulation signal is applied, the accuracy deteriorates rapidly, so it is necessary to use an approximation formula including the higher order term.

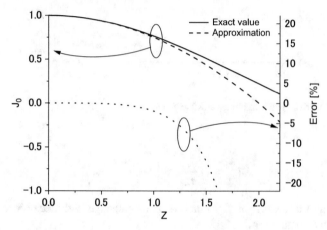

Fig. 6.2 Comparison between the exact value and approximation for J_0

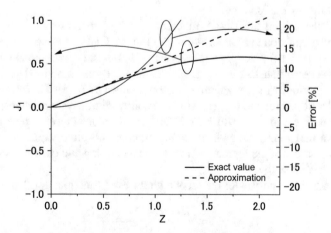

Fig. 6.3 Comparison between the exact value and approximation for J_1

The Bessel's differential equations represent the radial components of the cylindrical wave. When $|z|$ is very large, the Bessel function $J_n(z)$ can be approximately expressed by the asymptotic function as follows:

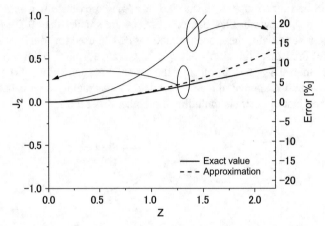

Fig. 6.4 Comparison between the exact value and approximation for J_2

$$J_n(z) \simeq \sqrt{\frac{2}{\pi z}} \cos \left(z - \frac{2n + 1}{4} \pi \right),$$ (6.18)

where the absolute value of $J_n(z)$ has oscillatory change with decay inversely proportional to the square root of $|z|$.

Figure 6.5 shows a three-dimensional surface created by $J_n(z)$ when z and n are changed near the origin and in positive real numbers, where the gamma function $\Gamma(n + 1) = n!$ is used to extend the factorial to a real number, so that n in (6.8) can be defined for any real number.

Around the origin, the Bessel functions ($n \neq 0$) increase with respect to z, smoothly in proportion to z^n as shown in (6.11), where $J_0(z)$ is a decreasing function as described in (6.12). Each Bessel function ($n \neq 0$) has a maximum at a particular z. When z is larger than that gives the maximum, the Bessel function has oscillatory change as described by the asymptotic function, shown in (6.18). This means that the higher order Bessel functions $J_n(z)$ will have non-negligible values when z becomes large, but each component will have a mixture of values close to zero and close to the local maximum, rather than having an equal absolute value. As shown in Fig. 6.5, the border of the regions of monotonic increasing and oscillatory change is approximately close to $n \sim z$.

Figure 6.6 shows curves for the second kind Bessel functions $Y_n(z)$ defined by

$$Y_n(z) = \lim_{\Omega \to n} \frac{J_n(z) \cos \Omega \pi - J_{-n}(z)}{\sin \Omega \pi}.$$ (6.19)

These Bessel functions that diverge to infinity at the origin are also solutions of the Bessel differential equation. The second kind Bessel functions $Y_n(z)$ are called the

Neumann functions, which can be described by

$$Y_n(z) \simeq \sqrt{\frac{2}{\pi z}} \sin\left(z - \frac{2n+1}{4}\pi\right) = J_{n+1}(z), \qquad (6.20)$$

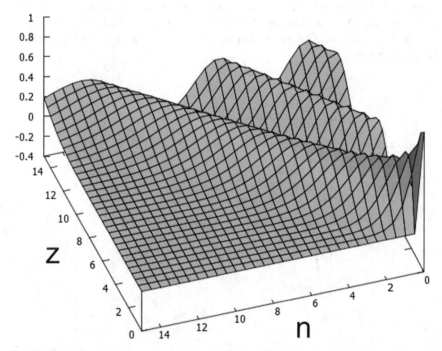

Fig. 6.5 Three-dimensional curvature created by the Bessel function when the order is continuously changed

approximately, when $|z|$ is large, as with the second kind Bessel functions.

When $|z| \sim 0$, the second kind Bessel function, $Y_n(z)$, has significant difference from the first kind Bessel function, whereas, far away from the origin, $J_n(z)$ and $Y_n(z)$ vary smoothly as well.

Equation (6.20) shows that $Y_n(z)$ is nearly equal to $J_{n+1}(z)$, where $J_n(z)$ and $Y_n(z)$ have $\pi/2$ phase difference in the oscillatory change. The modified Bessel function is defined from the Bessel function when z is a pure imaginary number, and the spherical Bessel function is defined from the Bessel function with the order of the Bessel function shifted by 1/2. These are the solutions of modified Bessel differential equations and spherical Bessel differential equations, respectively.

6.1.3 Definition of Bessel Function by Generation Function

In the field of applied mathematics, it is often explained that the Bessel function is defined as a solution to the Bessel differential equations as described earlier, but the relationship with phase modulation is not so clear. In the following, various mathematical expressions useful for the mathematical representation of phase modulation are derived from the definition by the generation function of the Bessel function.

The generation function of the Bessel functions whose order is 0 or positive integers is given by

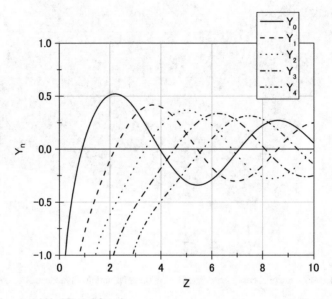

Fig. 6.6 The second kind Bessel function

$$e^{\frac{z}{2}(t-t^{-1})} = \sum_{n=0}^{\infty} [t^n + (-1)^n t^{-n}] J_n(z) - J_0(z). \tag{6.21}$$

As with the case of (6.8), the Bessel functions for negative integers are defined by

$$J_{-n}(z) = (-1)^n J_n(z), \tag{6.22}$$

so that the generation function can be rewritten as

$$e^{\frac{z}{2}(t-t^{-1})} = \sum_{n=-\infty}^{\infty} t^n J_n(z). \tag{6.23}$$

This function can be used as a generation function for the Bessel function $J_n(z)$ with integer orders including negative numbers. The generation function can be rewritten by using Taylor's expansion as follows:

$$\begin{aligned}
e^{\frac{z}{2}(t-t^{-1})} &= e^{\frac{zt}{2}} e^{-\frac{z}{2t}} \\
&= \left[\sum_{l=0}^{\infty} \left(\frac{zt}{2}\right)^l \frac{1}{l!}\right]\left[\sum_{m=0}^{\infty} \left(\frac{-z}{2t}\right)^m \frac{1}{m!}\right],
\end{aligned} \tag{6.24}$$

where the coefficient of the term of t^n given by the sum of the elements for which $l = m + n$ is expressed by

$$\sum_{m=0}^{\infty} \frac{(-1)^m}{m!(n+m)!} \left(\frac{z}{2}\right)^{n+2m} = J_n(z). \tag{6.25}$$

Thus, $J_n(z)$ defined by this generation function is equal to that of (6.8).

Here, n is assumed to be a non-negative integer; however, if n is a negative integer, the coefficient of the term of t^{-n} given by the sum of the elements for which $m = l - n$ is expressed by

$$\begin{aligned}
\sum_{l=0}^{\infty} \frac{(-1)^{l-n}}{l!(-n+l)!} \left(\frac{z}{2}\right)^{-n+2l} &= (-1)^{-n} \sum_{l=0}^{\infty} \frac{(-1)^l}{l!(-n+l)!} \left(\frac{z}{2}\right)^{-n+2l} \\
&= (-1)^{-n} J_{-n}(z).
\end{aligned} \tag{6.26}$$

Thus, by using $J_{-n}(z) = (-1)^n J_n(z)$, the definition of the Bessel functions by the generation function can be applied to the whole integer including negative numbers.

The generation function is invariant even if t is replaced by $-t^{-1}$ in the left-hand side, and hence we get

$$\begin{aligned}
e^{\frac{z}{2}(t-t^{-1})} &= \sum_{n=-\infty}^{\infty} (-t^{-1})^n J_n(z) \\
&= \sum_{n=-\infty}^{\infty} (-t^{-n}) J_n(z) \\
&= \sum_{n=-\infty}^{\infty} t^{-n}(-1)^n J_n(z).
\end{aligned} \tag{6.27}$$

By comparing this equation with (6.23), the definition of the Bessel functions for negative integers is as follows:

$$J_{-n}(z) = (-1)^n J_n(z). \tag{6.28}$$

The Bessel function is an even function when the order n is an even number and an odd function when the order n is an odd number.

The derivative of the generation function with respect to z is given by

$$\frac{1}{2}\left(t - t^{-1}\right) \sum_{n=-\infty}^{\infty} t^n J_n(z) = \sum_{n=-\infty}^{\infty} t^n \frac{dJ_n(z)}{dz}. \tag{6.29}$$

By using the coefficient comparison for t^n,

$$J_{n-1}(z) - J_{n+1}(z) = 2\frac{d}{dz}J_n(z) \tag{6.30}$$

can be obtained. Similarly, if we derive the generation function at t, we get

$$\frac{z}{2}\left(1 + t^{-2}\right) \sum_{n=-\infty}^{\infty} t^n J_n(z) = \sum_{n=-\infty}^{\infty} n t^{n-1} J_n(z). \tag{6.31}$$

By comparing the terms in t^{n-1} with the coefficients,

$$J_{n-1}(z) + J_{n+1}(z) = \frac{2n}{z} J_n(z) \tag{6.32}$$

can be obtained. The sum and difference of (6.30) and (6.32) provide the recurrence between Bessel functions where the orders have difference of 1 as follows:

$$J_{n-1}(z) = \frac{d}{dz}J_n(z) + \frac{n}{z}J_n(z) \tag{6.33}$$

$$J_{n+1}(z) = -\frac{d}{dz}J_n(z) + \frac{n}{z}J_n(z). \tag{6.34}$$

By substituting (6.33) to the equation obtained by reducing the order ($n \rightarrow n-1$) in (6.34), we get

$$\begin{aligned}
J_n(z) &= -\left[J_n'(z) + \frac{n}{z}J_n(z)\right]' + \frac{n-1}{z}\left[J_n'(z) + \frac{n}{z}J_n(z)\right] \\
&= -J_n''(z) + \frac{n}{z^2}J_n(z) - \frac{n}{z}J_n'(z) \\
&\quad + \frac{n-1}{z}J_n'(z) + \frac{n(n-1)}{z^2}J_n(z) \\
&= -J_n''(z) - \frac{1}{z}J_n'(z) + \frac{n^2}{z^2}J_n(z). \tag{6.35}
\end{aligned}$$

This is rewritten as

$$J_n''(z) + \frac{1}{z}J_n'(z) + \left(1 - \frac{n^2}{z^2}\right)J_n(z) = 0, \tag{6.36}$$

where $J_n'(z) = dJ_n(z)/dz$. This means that the Bessel functions $J_n(z)$ defined by the generation function Eq. (6.23) are solutions of the Bessel differential equations, Eq. (6.9).

6.1.4 Expansion of Composite Trigonometric Functions by Bessel Functions

By substituting $t = e^{i\theta}$ for the generating function of the Bessel function defined by (6.21), we get

$$
\begin{aligned}
e^{iz\sin\theta} = J_0(z) &+ 2iJ_1(z)\sin\theta + 2J_2(z)\cos 2\theta \\
&+ 2iJ_3(z)\sin 3\theta + 2J_4(z)\cos 4\theta \cdots,
\end{aligned}
\tag{6.37}
$$

where the real and imaginary components are, respectively,

$$
\cos(z\sin\theta) = J_0(z) + 2\sum_{n=1}^{\infty} J_{2n}(z)\cos(2n\theta)
\tag{6.38}
$$

$$
\sin(z\sin\theta) = 2\sum_{n=1}^{\infty} J_{2n+1}(z)\sin\{(2n+1)\theta\}.
\tag{6.39}
$$

By replacing θ with $\pi/2 - \theta$, these two equations can be rewritten as

$$
\cos(z\cos\theta) = J_0(z) + 2\sum_{n=1}^{\infty}(-1)^n J_{2n}(z)\cos(2n\theta)
\tag{6.40}
$$

$$
\sin(z\cos\theta) = 2\sum_{n=1}^{\infty}(-1)^n J_{2n+1}(z)\cos\{(2n+1)\theta\},
\tag{6.41}
$$

where we used the following equations:

$$
\begin{aligned}
\cos\left[2n(\pi/2 - \theta)\right] &= \cos\left[n\pi - 2n\theta\right] \\
&= \cos\left[2n\theta - n\pi\right]
\end{aligned}
\tag{6.42}
$$

$$
\begin{aligned}
\sin\left[(2n+1)(\pi/2 - \theta)\right] &= \sin\left[\pi/2 - (2n+1)\theta + n\pi\right] \\
&= (-1)^n \sin\left[\pi/2 - (2n+1)\theta\right] \\
&= (-1)^n \cos\left[(2n+1)\theta\right].
\end{aligned}
\tag{6.43}
$$

Equations (6.38)–(6.41) mean that composite trigonometric functions, i.e., $\cos(z\sin\theta)$, $\sin(z\sin\theta)$, $\cos(z\cos\theta)$, and $\sin(z\cos\theta)$, can be expressed by expansions with the Bessel functions, which are called the Jacobi–Anger expansions.

By using the addition theorem of trigonometric functions:

$$
\cos(A + B) = \cos A \cos B - \sin A \sin B,
\tag{6.44}
$$

with the Jacobi–Anger expansions shown in (6.38) and (6.40), we can derive the following equations:

$$\cos(z \sin \theta + \beta) = \sum_{n=-\infty}^{\infty} J_n(z) \cos(\beta + n\theta), \tag{6.45}$$

$$\cos(z \cos \theta + \beta) = J_0(z) \cos \beta$$

$$+ \sum_{n=1}^{\infty} (-1)^n J_{2n}(z) \{\cos(\beta + 2n\theta) + \cos(\beta - 2n\theta)\}$$

$$- \sum_{n=1}^{\infty} (-1)^n J_{2n+1}(z) \{\sin[\beta + (2n+1)\theta]$$

$$+ \sin[\beta - (2n+1)\theta]\} . \tag{6.46}$$

By using

$$\sin(A + B) = \sin A \cos B + \cos A \sin B, \tag{6.47}$$

for (6.38) and (6.40), we get

$$\sin(z \sin \theta + \beta) = \sum_{n=-\infty}^{\infty} J_n(z) \sin(\beta + n\theta) \tag{6.48}$$

$$\sin(z \cos \theta + \beta) = \cos \beta \sin(z \cos \theta) + \sin \beta \cos(z \cos \theta)$$

$$= J_0(z) \sin \beta$$

$$+ \sum_{n=1}^{\infty} (-1)^n J_{2n}(z) \{\sin[\beta + 2n\theta]$$

$$+ \sin[\beta - 2n\theta]\}$$

$$- \sum_{n=1}^{\infty} (-1)^n J_{2n+1}(z) \{\cos[\beta + (2n+1)\theta]$$

$$+ \cos[\beta - (2n+1)\theta]\} . \tag{6.49}$$

By using these equations, we can derive expansions for the phasor composite functions as follows:

$$\exp[i(z \sin \theta + \beta)] = \sum_{n=-\infty}^{\infty} J_n(z) \exp[i(\beta + n\theta)] \tag{6.50}$$

$$\exp[i(z \cos \theta + \beta)] = \sum_{n=-\infty}^{\infty} J_n(z) i^n \exp[i(\beta + n\theta)]. \tag{6.51}$$

The detail derivation of these equations: (6.45)–(6.51) is provided in Appendix A.

In short, the equations required for the expansion of the phase modulated signal described in (6.7) are

$$e^{i(z\sin\theta+\beta)} = e^{i\beta} \sum_{n=-\infty}^{\infty} J_n(z)e^{in\theta} \tag{6.52}$$

$$e^{i(z\cos\theta+\beta)} = e^{i\beta} \sum_{n=-\infty}^{\infty} J_n(z)i^n e^{in\theta}. \tag{6.53}$$

By multiplying with i^{-n} on the both sides of (6.28), we get

$$J_{-n}(z) = (-1)^n J_n(z)$$

$$= (i^2)^n J_n(z) \tag{6.54}$$

$$J_n(z)i^n = J_{-n}(z)i^{-n}, \tag{6.55}$$

so that

$$\sum_{n=-\infty}^{\infty} J_n(z)i^n e^{in\theta} = \sum_{n=-\infty}^{\infty} J_n(z)i^n e^{-in\theta}. \tag{6.56}$$

The expansions with the Bessel function for the phase modulation with phasor, i.e., (6.52) and (6.53) can be directly derived from the generation function as follows:

$$2i\sin\theta = e^{i\theta} - e^{-i\theta} \tag{6.57}$$

$$2\cos\theta = e^{i\theta} + e^{-i\theta} \tag{6.58}$$

to the right-hand side of (6.37), we get

$$e^{iz\sin\theta} = J_0(z) + [e^{i\theta} - e^{-i\theta}]J_1(z) + [e^{2i\theta} + e^{-2i\theta}]J_2(z)$$
$$+ [e^{3i\theta} - e^{-3i\theta}]J_3(z) + [e^{4i\theta} + e^{-4i4\theta}]J_4(z)\cdots. \tag{6.59}$$

From (6.28), we get $J_n(z) = J_{-n}(z)$ when n is an even number, and $J_n(z) = -J_{-n}(z)$ when n is an odd number. Thus, we can obtain

$$e^{iz\sin\theta} = \sum_{n=-\infty}^{\infty} J_n(z)e^{in\theta}, \tag{6.60}$$

which is identical to (6.52).

6.2 Sideband Generation by Phase Modulation

As mentioned in the previous section, the phase modulation by sinusoidal signals can be expanded by the Bessel functions. This is a theoretical explanation of sideband generation by modulation, and it is a basis for the expansion of the modulation signal in the frequency domain. Here, we describe the properties of the sideband generated by the phase modulation. Furthermore, we show that the optical modulation is linear to the optical input, and phase modulation for multiple spectral components can be considered as the optical input with a single spectral component.

We also discuss the physical interpretation of the addition theorem of the Bessel functions, which describe sideband generation by cascaded optical phase modulators. The output signals can be expressed as a sum of the sideband components generated by individual modulators. The addition theorem shows that the sidebands generated by the integrated modulator whose length is equal to the total length of cascaded modulators are consistent with the sideband components generated by the cascaded modulators.

6.2.1 Expansion of Phase Modulation by Bessel Functions

By using (6.52), we can rewrite (6.3) as follows:

$$
\begin{aligned}
R_i &= K_i e^{i\{\omega_0 t + iA_i \sin(\omega_m t + \phi_i) + B_i\}} \\
&= K_i e^{i\omega_0 t} e^{iA_i \sin(\omega_m t + \phi_i)} e^{iB_i} \\
&= K_i e^{i\omega_0 t} e^{iB_i} \sum_{n=-\infty}^{\infty} \left[J_n(A_i) e^{in\omega_m t + in\phi_i} \right] \\
&= K_i \sum_{n=-\infty}^{\infty} \left[J_n(A_i) e^{i(\omega_0 + n\omega_m)t + in\phi_i + iB_i} \right].
\end{aligned}
\tag{6.61}
$$

By omitting the constant amplitude and phase coefficients K_i and B_i, we get

$$
r_i = e^{iA_i \sin(\omega_m t + \phi_i)} = \sum_{n=-\infty}^{\infty} \left[J_n(A_i) e^{in\omega_m t + in\phi_i} \right].
\tag{6.62}
$$

The output (modulated) signal can be expressed by the product of this equation and the optical input, $e^{i\omega_0 t}$. This formula is most important as a mathematical representation of phase modulation, and the n-th term shows that the modulation generates a component whose angular frequency is shifted by $n\omega_m$, where the amplitude is given by $J_n(A_i)$. The n-th term is called the sideband whose frequency is $(\omega_0 + n\omega_m)/2\pi$. The zeroth order term whose frequency is identical to that of the input optical signal is called the carrier component. In this book, we call this component the zeroth order sideband, as well. The component whose frequency is higher than the carrier frequency, i.e., the component where n corresponds to the order of a positive integer, is called the upper sideband (USB). If only the relative phase is discussed, the time at which the phase difference between the zero order and the first order components is zero shall be used as reference. The negative component of n is called the lower sideband (LSB).

The sideband components with $\phi_i = 0$ are shown in Fig. 6.7. Since each sideband component vibrates at a different frequency, the phase relationship between each other differs depending on the time. Unless otherwise noted, the absolute phase is defined for the zeroth sideband component. When only the relative phase is needed, the time at which the phase difference between the zero order and the first order

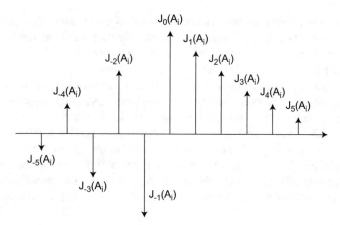

Fig. 6.7 Spectrum of phase modulated signal

components is zero shall be used as reference. Figure 6.7 shows the amplitude of each sideband component at a time where the absolute phase of the zeroth component is zero and the phase difference of the zeroth and first components is zero. The USB components ($n > 0$) are aligned in the same phase (0-degree), while the signs of the LSB components are inverted at even order (0-degree) and odd order sidebands (180-degree).

Here, we discuss the phase rotation of each sideband component, by using a three-dimensional expression of the sideband component phases, where the rotation around the horizontal axis (frequency axis) shows the phase. Regarding the horizontal axis, the rotation of the right screw direction will show that the phase leads.

Figure 6.8 shows the sideband components up to the second order ($n = -2, -1, 0, +1, +2$), corresponding to a part of the sidebands shown in Fig. 6.7 where the absolute phase for the 0th-order component is zero and the phase difference between the 0th-order and the 1st order components is also zero (which is the reference of the time axis: $t = 0$). By shifting the optical phase of the input light or the overall optical phase of the output light component by $\pi/2$, all components are rotated 90 degrees around the horizontal axis as shown in Fig. 6.9. There is no change in the relative phase relationship between the components.

Figure 6.10 shows the sideband components in the optical spectrum of the phase modulated signal with the $\pi/2$ phase shift of the modulating signal, where the phases of the sideband components are described as the rotation angles along the horizontal axis as with Fig. 6.7. Thus, the angular frequency of the modulating signal is ω_m, and the time delay corresponding to the $\pi/2$ phase shift is $t = \pi/2\omega_m$. In other words, the phase differences between the sideband components shown in Fig. 6.10 correspond to the phase differences after a quarter of the period from the reference time $t = 0$, where the phases of the sidebands at the reference time are shown in Fig. 6.7.

The first order USB component rotates in the right screw direction (positive direction) around the horizontal (frequency) axis with the angular velocity of ω_m.

The second order USB component rotates with the angular velocity $2\omega_m$. In general, the angular velocity of the n-th order USB is $n\omega_m$. On the other hand, the first and the second order LSB components rotate in the left screw direction (negative direction) with the angular velocities of ω_m and $2\omega_m$, respectively.

If the polarity of the angular velocity describes the rotation direction, the n-th order sideband component where n is an integer (including negative number) rotates along the horizontal axis with the angular velocity of $n\omega_m$. The phases of the sideband components described in Figs. 6.7, 6.8, 6.9, and 6.10 use the time domain reference corresponding to the time evolution term of the zeroth order component, i.e., $e^{i\omega_0 t}$, where the zeroth order component is fixed. If we use the time evolution term of the first LSB ($n = -1$) sideband for the reference, the zeroth and the first USB components rotate with the angular velocities of ω_m and $2\omega_m$, respectively.

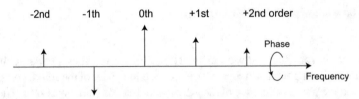

Fig. 6.8 Spectrum of phase modulated signal ($t = 0$)

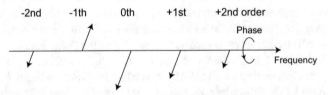

Fig. 6.9 Spectrum of phase modulated signal with $\pi/2$ phase shift of the optical input

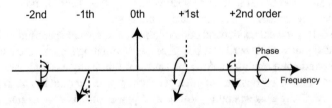

Fig. 6.10 Spectrum of phase modulated signal with $\pi/2$ phase shift of the modulating signal

As increasing the amplitude of the modulating signal A_i from zero, J_0 decreases as a quadratic function shown in (6.17). $J_1, J_2, J_3 \cdots$, respectively, increase in proportion to $z, z^2, z^3 \cdots$. J_1 decreases after the maximum of 0.582 at $z = 1.84$ and fluctuates with oscillatory changes described by (6.18).

The Bessel function for each n ($n \neq 0$) has the maximum where $dJ_n/dz = 0$ has the solution closest to zero. From (6.30), the derivative of J_n with respect to z can be expressed by

$$\frac{dJ_n(z)}{dz} = \frac{1}{2}\left(J_{n-1}(z) - J_{n+1}(z)\right). \tag{6.63}$$

For the zeroth order component, we get

$$\frac{dJ_0(z)}{dz} = -J_1(z). \tag{6.64}$$

Thus, the derivative of the Bessel functions can be easily calculated from the numerical values of the Bessel functions. Table 6.1 shows values of z ($z > 0$) that give the maxima and zero closest to $z = 0$. The required phase shift for OOK is 0.8, and that for 180-degree phase shift is about 1.6 in radian. These values are smaller than 1.84 where J_1 has the maximum.

Under the common operating conditions of the modulator, where $|z|$ would be less than 2, J_n is rarely subject to vibration, and in particular, the approximation that the high order sideband component where n is greater than or equal to 2 increases monotonically in proportion to z^n is useful.

Table 6.1 Maxima and zeros of the Bessel functions

	J_0	J_1	J_2	J_3	J_4	J_5
Maxima	1	0.582	0.486	0.434	0.400	0.374
z for maxima	0	1.84	3.05	4.20	5.32	6.42
z for zeros	2.40	3.83	5.14	6.38	7.59	8.77

The initial phase of the modulating signal ϕ_i shifts the optical phase of each sideband with $n\phi_i$. The wavelength of the electric modulating signal is 30 mm (in vacuum) for 10 GHz, while that of the optical signal is 1.5 μm. It is an interesting phenomenon that the phase of the electric signal is directly connected to the change of the optical phase, even though there is a difference of about 20,000 times in the physical length of the wavelengths. As shown in Fig. 4.42, in general, devices whose sizes are comparable to the wavelengths of the signals are required to control the phases of the signals, so that it is rather difficult to control low-frequency components. Such phase differences associated with large device structures have direct connection to the optical phases with small wavelengths.

6.2.2 Power Conservation in Phase Modulation

Here, we consider the optical power of the phase modulated signal shown in Fig. 6.7. By using (6.61), optical power of each sideband component can be derived as follows:

$$P_n = \frac{K_i}{Z_c} \{J_n(A_i)\}^2, \tag{6.65}$$

where K_i denotes the amplitude of the optical input. When the optical loss in the optical waveguide is not negligible, K_i should include the decay due to the loss. Z_c is a coefficient associated with the characteristic impedance of the medium. For simplicity, we assume that $Z_c = 1$ in this book. Phase modulation does not change the optical power, so that the sum of power of all the sideband components should be equal to the input optical power. By assuming that $K_i = 1$ without loss of generality, the total power of the phase modulated signal should satisfy the following equation:

$$P_t = \sum_{n=-\infty}^{\infty} \{J_n(A)\}^2 = 1, \tag{6.66}$$

for an arbitrary real number A (see Problem 6.2). By using

$$\{J_n(A_i)\}^2 = \{J_{-n}(A_i)\}^2, \tag{6.67}$$

this equation can be rewritten as

$$P_t = \{J_0(A)\}^2 + 2\sum_{n=1}^{\infty} \{J_n(A)\}^2 = 1. \tag{6.68}$$

This equation means that the optical input power is distributed into sideband components, where the energy conservation law is satisfied.

In numerical calculations, we can include only a limited number of terms. Bessel functions are also approximately described by polynomials, as shown in (6.8). In (6.13), (6.13), and (6.17), $J_0(A)$, $J_1(A)$, and $J_2(A)$ can be expressed by second-degree polynomials of A. The sum of the sideband components is

$$P_{t2}^{(2)} = \{J_0(A)\}^2 + 2\{J_1(A)\}^2 + 2\{J_2(A)\}^2 = 1 + \frac{3}{32}A^4. \tag{6.69}$$

The difference defined by

$$\frac{P_{t2}^{(2)} - P_t}{P_t} = \frac{3}{32}A^4 \tag{6.70}$$

reflects numerical calculation error in this approximation. This difference can be useful to estimate accuracy of the approximation.

Similarly, the sum of the sideband components up to the fourth order, approximated by fourth-degree polynomials of A (see Problem 6.4), is given by

$$P_{t4}^{(4)} = 1 - \frac{5A^6}{1152} + \frac{35A^8}{73728}. \tag{6.71}$$

Figures 6.11 and 6.12 show the sum of the sideband components up to the second order, $P_{t2}^{(2)}$, and that for up to the fourth order sidebands, $P_{t4}^{(4)}$, as functions of the induced optical phase shift A. When A is less than 0.5 rad, the power difference from the exact value is less than 1% both in $P_{t2}^{(2)}$ and $P_{t4}^{(4)}$. On the other hand, the difference of $P_{t2}^{(2)}$ rapidly increases with A, when $A > 1$.

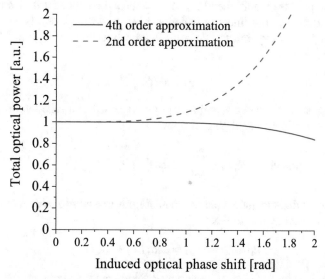

Fig. 6.11 Sums of powers in sideband components up to finite orders. Dashed and solid lines show $P_{t2}^{(2)}$ and $P_{t4}^{(4)}$, respectively

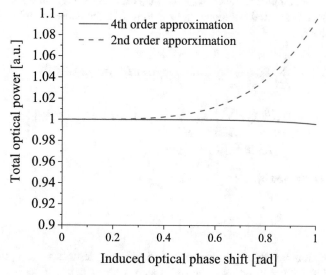

Fig. 6.12 Blow-up of Fig. 6.11

6.2.3 Linearity in Optical Phase Modulation

The optical modulation by the EO effect, i.e., the Pockels effect, is linear to the optical input when the higher order effects such as the Kerr effect can be neglected. Thus, if the input amplitude is multiplied by x, each sideband component is also multiplied by x proportionally. The linearity is degraded if the transmission is dependent on the input light intensity, but the linearity is known to be very high in LN and LT modulators.

In (6.61),

$$K_i = K_{Li} E_0, \tag{6.72}$$

so that the phase modulated signal R_i can be expressed as a function of a complex amplitude of the input signal E_0 as follows:

$$R_i(E_0) = E_0 K_{Li} \sum_{n=-\infty}^{\infty} \left[J_n(A_i) e^{i(\omega_0 + n\omega_m)t + in\phi_i + iB_i} \right]. \tag{6.73}$$

It is evident that the optical phase modulation is linear with respect to the amplitude of the optical inputs, so that

$$R_i(aE_a + bE_b) = aR_i(E_a) + bR_i(E_b). \tag{6.74}$$

Here we consider an optical input consisting of multiple components described by

$$\sum_{k=1}^{m} E_{0k} e^{i\omega_{0k}t} = \sum_{k=1}^{m} |E_{0k}| e^{i(\omega_{0k}t + \phi_{0k})}, \tag{6.75}$$

where E_{0k} is a complex number. $|E_{0k}|$ and ϕ_k are, respectively, the amplitude and phase of the k-th spectral component. As with (6.61), the phase modulated signal is expressed by

$$\sum_{k=1}^{m} E_{0k} K_{Li} \exp \left[i \{ \omega_{0k} t + A_i \sin(\omega_m t + \phi_i) + B_i \} \right]$$

$$= e^{iB_i} K_{Li} \sum_{k=1}^{m} \left[E_{0k} \sum_{n=-\infty}^{\infty} \left\{ J_n(A_i) e^{i(\omega_{0k} + n\omega_m)t + in\phi_i} \right\} \right]$$

$$= E_{01} R_{i1} + E_{02} R_{i2} + \cdots E_{0m} R_{im}, \tag{6.76}$$

where R_{ik} is denoted by

$$R_{ik} = e^{iB_i} K_{Li} \sum_{n=-\infty}^{\infty} \left\{ J_n(A_i) e^{i(\omega_{0k} + n\omega_m)t + in\phi_i} \right\}. \tag{6.77}$$

This expression for R_{ik} is derived from (6.61) and describes the optical phase modulated signal produced from the k-th spectral component with unit amplitude.

Equation (6.76) means that the phase modulated light for an input signal consisting of multiple spectrum components can be expressed by linear coupling of R_{ik}, which is the phase modulated light with only each component being the input light.

Figure 6.13 shows the features of the linearity of the optical phase modulation. The optical output for optical input $aQ + bQ'$ is given by $aR + bR'$, where R and R' denote the optical outputs for Q and Q', respectively. When the inputs of Q and Q' have the same frequency, the output is expressed by (6.74). When Q and Q' have different frequencies, Eq. (6.76) gives the output. As described earlier, the optical phase modulation has the wide range of linearity for the optical input; however, the modulation process is nonlinear for the modulation signal whose amplitude is denoted by A_i as shown in (6.61). This is why the complex numbers can be used for the optical signal expression, whereas the real numbers should be used for the modulation signals, which are electrical signals.

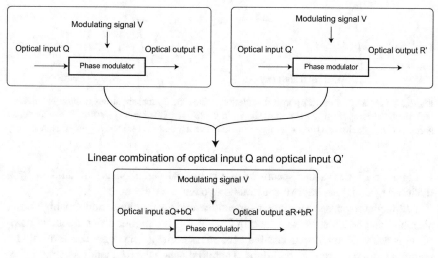

Fig. 6.13 Linearity of optical phase modulation

Here, consider the case where the input light is modulated beforehand with an information signal, etc., and has a spectrum with a finite bandwidth. The input light can be decomposed into the individual optical frequency components by Fourier transform. Since the above linearity holds for each optical frequency component, as shown in Fig. 6.14, the output of the optical phase modulator has components in which the spectrum of the input light is replicated where the center frequency deviates by an integral multiple of the modulation frequency.

This property in the sideband generation by optical modulation can be applied to optical frequency conversion and signal processing. Various modulators based on optical phase modulators utilize this property to realize functions to shift the center frequency and to generate a duplicate spectrum with different center frequency without losing the waveform information of the input light.

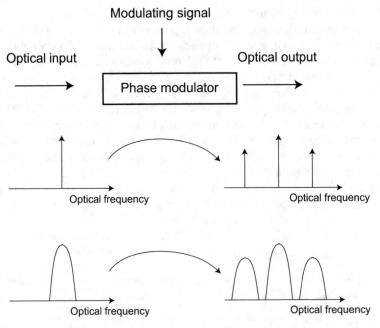

Fig. 6.14 Linearity of optical phase modulation. When the optical input has a finite bandwidth, each sideband component has a spectral shape similar to that of the input spectrum. The vertical and horizontal axes of the spectrum diagram show the intensity and the optical frequency, respectively

Figure 6.15 shows the spectrum of the optical signal phase modulated by a modulating signal having two spectral components: $e^{i\omega_{01}t}$ and $e^{i\omega_{02}t}$.

Although the phase modulation signal theoretically contains infinitely high order sideband components, as explained in Sect. 6.2.1, the higher order sideband components are very small and less than the measurement limit or noise level, under commonly used operating conditions. Therefore, when the frequency separation between the two spectral components of $e^{i\omega_{01}t}$ and $e^{i\omega_{02}t}$ is large enough to be more than a multiple of the modulating frequency as shown in Fig, 6.15a, the sidebands generated from the respective two spectral components are far apart on the optical frequency axis and do not interfere with each other. The optical modulation signal is eventually converted into an electrical signal by a photodetector, but the optical frequency component that is further apart than the response frequency of the optical detector does not affect the electrical signal.

As shown in Fig. 6.15b, beat signals are generated at the photodetector, when some of the sideband components, such as those enclosed in dotted lines, are close enough in the frequency domain. The beat signal frequency corresponds to the frequency difference between the sideband components.

Figure 6.15c shows the case where some of the sideband components from different components in the input match together in the frequency domain. The interference results in the addition or subtraction of the sideband components on the complex plane. If two independent laser sources are used as inputs, the interference shown in Fig. 6.15c would be unstable because the frequency of each source varies in time. In general, the relationship between the sideband components would be as shown in Fig. 6.15b, due to phase and frequency fluctuation of the laser sources. On the other hand, when more than one spectral component consisting of sideband components generated by optical modulation is used as the optical input, the optical frequency separation between the spectral components is very stable, so the effect of the interference shown in Fig. 6.15c can be obtained stably. Various sideband generation techniques with a highly stable configuration, in which the modulation signals shown in Fig. 6.16 have been proposed for optical signal generation and processing [1].

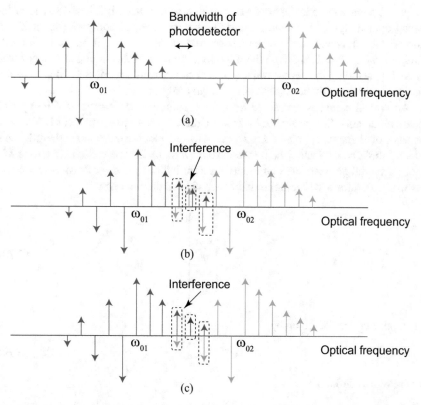

Fig. 6.15 Optical spectrum of phase modulated signal with two spectral components in the input. (**a**) the optical frequency interval is sufficiently larger than the modulation frequency and there is no overlap in the sideband component, (**b**) the sideband components overlap, and (**c**) some of the sidebands match in the frequency domain

The vector and matrix representations are useful for the analysis of cascaded modulation schemes, as shown in Fig. 6.16, where multiple sideband components are fed to additional modulators. In principle, the phase modulation generates high

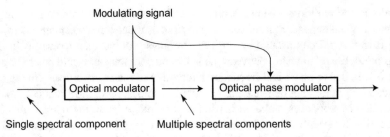

Fig. 6.16 Configuration of optical phase modulation for multi-spectrum components generated by optical modulation

order sideband components; however, as mentioned in Sect. 6.2.1, we can describe actual optical signals by using a finite number of sideband components. Here, we consider the sideband components whose order is up to n, i.e., from $-n$ (n-th order LSB) to $+n$ (n-th order LSB), where the total number of the sideband components is $2n + 1$ including the zeroth order (carrier) component. Of course, when $n \to \infty$, this is consistent with the rigorous theoretical analysis.

An optical signal is expressed by a $2n + 1$-dimensional vector $\boldsymbol{a}$. The optical phase modulation is represented by a $2n + 1$-dimensional square matrix $\boldsymbol{M}$. Vector $\boldsymbol{a}$ consists of elements a_k, where k is from $-n$ to $+n$ to let the index show the sideband order. Similarly, let j and k be integers from $-n$ to $+n$ for every element M_{jk} of $\boldsymbol{M}$, though natural numbers are commonly used for indices j and k of matrix elements. By using a vector $\boldsymbol{a}$ with amplitude of each sideband component

$$\boldsymbol{a} \equiv \begin{bmatrix} \vdots \\ a_{-1} \\ a_0 \\ a_{+1} \\ \vdots \end{bmatrix}, \tag{6.78}$$

the input light can be given by

$$Q = \mathrm{e}^{\mathrm{i}\omega_0 t}\, \boldsymbol{e}(\omega_m) \cdot \boldsymbol{a}, \tag{6.79}$$

where $\boldsymbol{e}(\omega_m)$ defined by

$$\boldsymbol{e}(\omega_m) \equiv \begin{bmatrix} \vdots \\ \cdots \\ \mathrm{e}^{-2\mathrm{i}\omega_m t} \\ \mathrm{e}^{-\mathrm{i}\omega_m t} \\ 1 \\ \mathrm{e}^{\mathrm{i}\omega_m t} \\ \mathrm{e}^{2\mathrm{i}\omega_m t} \\ \vdots \end{bmatrix} \tag{6.80}$$

describes time evolution of each element. The output of the modulator can be expressed by

$$R_i = e^{iB_i} e^{i\omega_0 t} K_i e(\omega_m) \cdot M(A_i, \phi_i) a. \tag{6.81}$$

$M(A_i, \phi_i)$, defined by

$$M(A_i, \phi_i) = \begin{bmatrix} \ddots & \vdots & \vdots & \vdots & \cdot \\ \cdots & M_{-1,-1}(A_i, \phi_i) & M_{-1,0}(A_i, \phi_i) & M_{-1,+1}(A_i, \phi_i) & \cdots \\ \cdots & M_{0,-1}(A_i, \phi_i) & M_{0,0}(A_i, \phi_i) & M_{0,+1}(A_i, \phi_i) & \cdots \\ \cdots & M_{+1,-1}(A_i, \phi_i) & M_{+1,0}(A_i, \phi_i) & M_{+1,+1}(A_i, \phi_i) & \cdots \\ \cdot & \vdots & \vdots & \vdots & \ddots \end{bmatrix} \tag{6.82}$$

describes operation of the optical phase modulator, where the element in the j-th row and k-th column is given by

$$M_{jk}(A_i, \phi_i) \equiv J_{(j-k)}(A_i) e^{i(j-k)\phi_i}, \tag{6.83}$$

which expresses the sideband generation for each input component. When the input is given by $a_0 = 1$ and $a_i = 0$ $(i \neq 0)$, the output is given by

$$M(A_i, \phi_i) a = \begin{bmatrix} \vdots \\ e^{-2i\phi_i} J_{-2}(A_i) \\ e^{-i\phi_i} J_{-1}(A_i) \\ J_0(A_i) \\ e^{i\phi_i} J_1(A_i) \\ e^{2i\phi_i} J_2(A_i) \\ \vdots \end{bmatrix}. \tag{6.84}$$

By substituting this for (6.81), we get

$$R_i = K_i \sum_{n=-\infty}^{\infty} \left[J_n(A_i) e^{i(\omega_0 + n\omega_m)t + in\phi_i + iB_i} \right], \tag{6.85}$$

which is identical to (6.61). Thus, we can deduce that the matrix expression described in this subsection extends the case of the single spectrum input ($E_0 e^{i\omega_0 t}$), shown in (6.61).

Similarly, assuming that $a_k = 1$ and $a_i = 0$ $(i \neq k)$, (6.81) becomes identical to (6.61) multiplied by $e^{ik\omega_m t}$, which corresponds to optical phase modulation for an optical input with a spectrum component associated with angular frequency $\omega_0 + k\omega_m t$.

6.2.4 Addition Theorem of Bessel Functions

In this section, we discuss the addition theorem of the Bessel functions related to phase modulation and their physical interpretation. By using the trigonometric addition theorem:

$$A_i \sin(\omega_m t + \phi_i) = A_i \sin \omega_m t \cos \phi_i + A_i \cos \omega_m t \sin \phi_i, \qquad (6.86)$$

(6.62) is rewritten as

$$
\begin{aligned}
e^{iA_i \sin(\omega_m t + \phi_i)} &= e^{i(A_i \sin \omega_m t \cos \phi_i + A_i \cos \omega_m t \sin \phi_i)} \\
&= e^{iA_i \sin \omega_m t \cos \phi_i} e^{iA_i \cos \omega_m t \sin \phi_i}, \qquad (6.87)
\end{aligned}
$$

which is a product of phase modulated signals with

$$A_i \sin \omega_m t \cos \phi_i,$$

and

$$A_i \cos \omega_m t \sin \phi_i,$$

in the phasor representation. By using (6.52) and (6.53), we can get an expansion of each phase modulation term with the Bessel functions as follows:

$$
\begin{aligned}
& e^{i(A_i \sin \omega_m t \cos \phi_i + A_i \cos \omega_m t \sin \phi_i)} \\
&= \sum_{n=-\infty}^{\infty} J_n(A_i \cos \phi_i) e^{in\omega_m t} \times \sum_{n=-\infty}^{\infty} J_n(A_i \sin \phi_i) i^n e^{in\omega_m t} \\
&= \sum_{n=-\infty}^{\infty} \sum_{s=-\infty}^{\infty} \{ J_{n-s}(A_i \cos \phi_i) J_s(A_i \sin \phi_i) e^{in\omega_m t} i^s \}. \qquad (6.88)
\end{aligned}
$$

By comparing the coefficient in the previous equation with the term of $e^{in\omega_m t}$ in (6.62), we get

$$J_n(A_i) e^{in\phi_i} = \sum_{s=-\infty}^{\infty} \{ J_{n-s}(A_i \cos \phi_i) J_s(A_i \sin \phi_i) i^s \}, \qquad (6.89)$$

which describes one of the addition theorems for the Bessel functions. This means that the same result can be obtained by separating a sinusoidal wave into two components with a trigonometric addition theorem and then expanding them with a Bessel function.

The generation function of the Bessel function can be rewritten with

$$\sum_{n=-\infty}^{\infty} J_n(x + y)t^n = e^{(1/2)(x+y)(t-1/t)}$$

$$= e^{(x/2)(t-1/t)}e^{(y/2)(t-1/t)}$$

$$= \sum_{s=-\infty}^{\infty} J_s(x)t^n \sum_{r=-\infty}^{\infty} J_r(y)t^n, \tag{6.90}$$

where Eq. (6.23) x is replaced with $x + y$.

By comparing the terms of t^n, we can obtain an addition theorem as follows:

$$J_n(x + y) = \sum_{s=-\infty}^{\infty} J_s(x)J_{n-s}(y). \tag{6.91}$$

This equation means that the sideband components generated by a modulating signal $(x + y)\sin \omega_m t$ can be described by the Bessel function expansion with the amplitude of $x + y$ identical to the product of the expansions with the amplitudes of x and y.

Figure 6.17 shows the configuration of the phase modulation with an optical phase modulator with a modulating signal with the amplitude of $x + y$ (zero-to-peak), where a modulating signal, $(x + y)\sin \omega_m t$, is fed to one modulator. For the sake of simplicity, the sign relationship of each sideband component is ignored in this figure, and only the intensity profile is shown as a conceptual diagram. The amplitude of the first order sideband is proportional to $|J_1(x + y)|$.

On the other hand, Fig. 6.18 shows the configuration of the cascaded optical phase modulation. A modulating signal of $y \sin \omega_m t$ is fed to one optical phase modulator, and the output of the modulator is connected to the optical input of another modulator to which a modulating signal of $x \sin \omega_m t$. The first modulator in Fig. 6.18 generates sidebands whose amplitudes are proportional to $|J_1(y)|$. The sideband components are further modulated at the second modulator, and the amplitudes of the sideband components are given by (6.91).

For example, the first order USB (+1-th order sideband) in the final optical output (the output of the second modulator) consists of components generated from various processes, such as the +2-th order sideband generated by the second modulator from −1-th order sideband generated by the first modulator, and 0-th order sideband generated by the second modulator from +1-th order sideband generated by the first modulator, where the frequency of the 0-th order sideband is identical to that of the input. The amplitude of the former process is $J_2(x)J_{-1}(y)$ and that of the latter process is $J_0(x)J_1(y)$. In general, the s-th order sideband generated at the second modulator from the $n - s$-th order sideband generated by the first modulator would contribute the n-th order sideband in the final output, where the amplitude is described by $J_s(x)J_{n-s}(y)$. As described in (6.91), the total amount of the n-th order sideband in the final output can be calculated by the sum of the sideband order in the output of the first modulator.

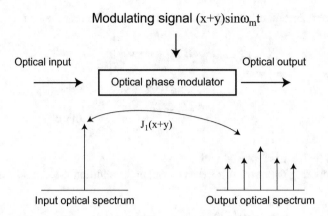

Fig. 6.17 Sideband generation by optical phase modulation with amplitude of $(x + y) \sin \omega_m t$

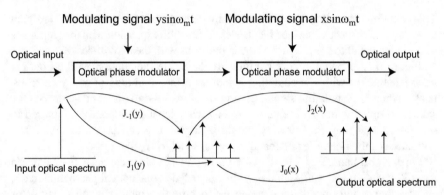

Fig. 6.18 Sideband generation by cascaded optical phase modulation processes with amplitudes of $y \sin \omega_m t$ and $x \sin \omega_m t$

Here, we consider an addition theorem for the matrix form of the optical phase modulation, $\boldsymbol{M}$:

$$\boldsymbol{M}(u,0) = \begin{bmatrix} \ddots & \vdots & \vdots & \vdots & \cdot \\ \cdots & J_0(u) & J_{-1}(u) & J_{-2}(u) & \cdots \\ \cdots & J_1(u) & J_0(u) & J_{-1}(u) & \cdots \\ \cdots & J_2(u) & J_1(u) & J_0(u) & \cdots \\ \cdot & \vdots & \vdots & \vdots & \ddots \end{bmatrix}, \tag{6.92}$$

which is defined by (6.82).

Assuming that the phase of the modulating signal is zero $\phi_i = 0$, as with the physical interpretation of (6.91), we discuss the cascaded sideband generation processes described by the product of the two matrices $\boldsymbol{M}(x, 0)$ and $\boldsymbol{M}(y, 0)$, which is given by

$$\boldsymbol{M}(x,0)\boldsymbol{M}(y,0) = \begin{bmatrix} \ddots & \vdots & \vdots & \vdots & \ddots \\ \cdots & J_0(x) & J_{-1}(x) & J_{-2}(x) & \cdots \\ \cdots & J_1(x) & J_0(x) & J_{-1}(x) & \cdots \\ \cdots & J_2(x) & J_1(x) & J_0(x) & \cdots \\ \cdot & \vdots & \vdots & \vdots & \ddots \end{bmatrix}$$

$$\times \begin{bmatrix} \ddots & \vdots & \vdots & \vdots & \cdot \\ \cdots & J_0(y) & J_{-1}(y) & J_{-2}(y) & \cdots \\ \cdots & J_1(y) & J_0(y) & J_{-1}(y) & \cdots \\ \cdots & J_2(y) & J_1(y) & J_0(y) & \cdots \\ \cdot & \vdots & \vdots & \vdots & \ddots \end{bmatrix}. \tag{6.93}$$

By using (6.91), the product can be rewritten as

$$\boldsymbol{M}(x,0)\boldsymbol{M}(y,0) = \begin{bmatrix} \ddots & \vdots & \vdots & \vdots & \cdot \\ \cdots & J_0(x+y) & J_{-1}(x+y) & J_{-2}(x+y) & \cdots \\ \cdots & J_1(x+y) & J_0(x+y) & J_{-1}(x+y) & \cdots \\ \cdots & J_2(x+y) & J_1(x+y) & J_0(x+y) & \cdots \\ \cdot & \vdots & \vdots & \vdots & \ddots \end{bmatrix}, \tag{6.94}$$

so that

$$\boldsymbol{M}(x+y,0) = \boldsymbol{M}(x,0)\boldsymbol{M}(y,0). \tag{6.95}$$

This means that the optical modulation of $(x+y)\sin\omega_m t$ is identical to the cascaded modulation of $y\sin\omega_m t$ and $x\sin\omega_m t$.

By using (6.95), iteratively, we get

$$\boldsymbol{M}(x,0) = \boldsymbol{M}(\Delta x_1,0)\boldsymbol{M}(\Delta x_2,0)\cdots\boldsymbol{M}(\Delta x_n,0), \tag{6.96}$$

for $\Delta x_1, \Delta x_2, \cdots, \Delta x_n$, where the modulation signal with the amplitude x is divided into small parts Δx_k, which are defined by

$$x = \sum_{k=1}^{n} \Delta x_k. \tag{6.97}$$

This means that the output of the phase modulation with the amplitude $x\sin\omega_m t$ is identical to that of the cascaded modulation processes with the amplitudes $\Delta x_1\sin\omega_m t, \Delta x_2\sin\omega_m t, \cdots, \Delta x_n\sin\omega_m t$.

In the earlier discussion, we neglected the impact of time delay between the modulation processes in the cascaded optical phase modulation. When the size of the modulation device is much smaller than the wavelength of the modulating signal in the device, the delay can be neglected in such analysis. However, the propagation delay of the lightwave and the modulating (electric) signal should be taken into consideration, when the device size is larger than a few cm or the modulation processes are performed by more than one discrete device connected by cables.

Figure 6.19 shows a schematic of the cascaded optical phase modulation process, where Δt_k and $\Delta t'_k$ denote propagation delays of the lightwave and the modulating (electric) signal between the k-th and $k + 1$-th modulation processes. At the k-th modulation process, the lightwave is modulated by a part of the modulating signal whose amplitude is $\Delta x_k \sin \omega_m t$. We should note that the modulating signal for the k-th modulation process has phase delay due to the sum of delays in the modulation processes: $\omega_m(\Delta t'_1 + \Delta t'_2 + \cdots + \Delta t'_{k-1})$.

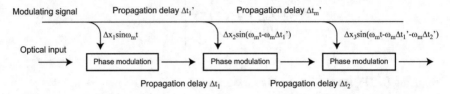

Fig. 6.19 Cascaded optical phase modulation processes with propagation delay effect

The sideband generation in the cascaded optical phase modulation can be expressed by

$$
\begin{aligned}
G = {} & T(-\Delta t_n)M(\Delta x_n, -\omega_m(\Delta t'_1 + \omega_m \Delta t'_2 + \cdots + \Delta t'_n)) \times \cdots \\
& \times T(-\Delta t_3)M(\Delta x_3, -\omega_m(\Delta t'_1 + \Delta t'_2)) \\
& \times T(-\Delta t_2)M(\Delta x_2, -\omega_m \Delta t'_1)T(-\Delta t_1)M(\Delta x_1, 0),
\end{aligned}
\tag{6.98}
$$

where the matrix $M(x, \phi)$ shows the effect of the optical phase modulation with the amplitude of $x \sin(\omega_m t + \phi)$, and the matrix $T(t)$ defined by

$$
T(t)
$$

$$
\equiv
\begin{bmatrix}
\ddots & \vdots & \vdots & \vdots & \vdots & \vdots & \ddots \\
\cdots & e^{i(\omega_0-2\omega_m)t} & 0 & 0 & 0 & 0 & \cdots \\
\cdots & 0 & e^{i(\omega_0-\omega_m)t} & 0 & 0 & 0 & \cdots \\
\cdots & 0 & 0 & 1 & 0 & 0 & \cdots \\
\cdots & 0 & 0 & 0 & e^{i(\omega_0+\omega_m)t} & 0 & \cdots \\
\cdots & 0 & 0 & 0 & 0 & e^{i(\omega_0+2\omega_m)t} & \cdots \\
\ddots & \vdots & \vdots & \vdots & \vdots & \vdots & \ddots
\end{bmatrix}
$$

$$
= e^{i\omega_0 t}
\begin{bmatrix}
\ddots & \vdots & \vdots & \vdots & \vdots & \vdots & \ddots \\
\cdots & e^{-2i\omega_m t} & 0 & 0 & 0 & 0 & \cdots \\
\cdots & 0 & e^{-i\omega_m t} & 0 & 0 & 0 & \cdots \\
\cdots & 0 & 0 & 1 & 0 & 0 & \cdots \\
\cdots & 0 & 0 & 0 & e^{i\omega_m t} & 0 & \cdots \\
\cdots & 0 & 0 & 0 & 0 & e^{2i\omega_m t} & \cdots \\
\ddots & \vdots & \vdots & \vdots & \vdots & \vdots & \ddots
\end{bmatrix}
\tag{6.99}
$$

describes the phase shifts of sideband components due to the time delay of t.

The component in the j-th row and k-th column of the matrix $\boldsymbol{T}(-\Delta t)$ that describes the phase shift due to the propagation delay Δt is given by

$$T_{jk}(-\Delta t) = \mathrm{e}^{-\mathrm{i}\omega_0 \Delta t} \delta_{jk} \mathrm{e}^{-\mathrm{i}k\omega_m \Delta t}, \tag{6.100}$$

where

$$\delta_{jk} = \begin{cases} 1 \ (j = k) \\ 0 \ (j \neq k) \end{cases}. \tag{6.101}$$

The components in the j-th row and k-th column of the matrices, $\boldsymbol{M}(\Delta x_1, 0)$ and $\boldsymbol{M}(\Delta x_2, -\omega_m \Delta t_1')$, are, respectively,

$$M_{jk}(\Delta x_1, 0) \equiv J_{(j-k)}(\Delta x_1) \tag{6.102}$$

and

$$M_{jk}(\Delta x_2, -\omega_m \Delta t_1') \equiv J_{(j-k)}(\Delta x_2) \mathrm{e}^{-\mathrm{i}(j-k)\omega_m \Delta t_1'}. \tag{6.103}$$

In general, the component in the j-th row and k-th column of the matrix, $\boldsymbol{M}(\Delta x_2, \omega_m \Delta t_1')\boldsymbol{T}(-\Delta t_1)$, is given by

$$\sum_{s=-n}^{n} J_{(j-s)}(\Delta x_2) \mathrm{e}^{-\mathrm{i}(j-s)\omega_m \Delta t_1'} \mathrm{e}^{-\mathrm{i}\omega_0 \Delta t_1} \delta_{sk} \mathrm{e}^{-\mathrm{i}k\omega_m \Delta t_1}$$
$$= \mathrm{e}^{-\mathrm{i}\omega_0 \Delta t_1} J_{(j-k)}(\Delta x_2) \mathrm{e}^{-\mathrm{i}j\omega_m \Delta t_1'} \mathrm{e}^{\mathrm{i}k\omega_m (\Delta t_1' - \Delta t_1)}. \tag{6.104}$$

By assuming that the propagation delay of the modulating signal is equal to that of the lightwave, i.e., $\Delta t_1 = \Delta t_1'$, we get

$$\mathrm{e}^{-\mathrm{i}\omega_0 \Delta t_1} \mathrm{e}^{-\mathrm{i}j\omega_m \Delta t_1} J_{(j-k)}(\Delta x_2) = \sum_{s=-n}^{n} \mathrm{e}^{-\mathrm{i}\omega_0 \Delta t_1} \delta_{js} \mathrm{e}^{-\mathrm{i}s\omega_m \Delta t_1} J_{(s-k)}(\Delta x_2), \tag{6.105}$$

which is identical to the component in the j-th row and k-th column of

$$\boldsymbol{T}(-\Delta t_1)\boldsymbol{M}(\Delta x_2, 0).$$

This means that the matrices $\boldsymbol{M}$ and $\boldsymbol{T}$ satisfy the commutation relation:

$$\boldsymbol{M}(\Delta x_2, -\omega_m \Delta t_1)\boldsymbol{T}(-\Delta t_1) = \boldsymbol{T}(-\Delta t_1)\boldsymbol{M}(\Delta x_2, 0) \tag{6.106}$$

when the modulating signal and the lightwave have the same propagation delay. By using the commutation relations, we get

$$
\begin{aligned}
\boldsymbol{G} &= \boldsymbol{T}(-\Delta t_n)\boldsymbol{M}(\Delta x_n, -\omega_m(\Delta t_1' + \Delta t_2' + \cdots + \Delta t_n')) \times \cdots \\
&\quad \times \boldsymbol{T}(-\Delta t_3)\boldsymbol{M}(\Delta x_3, -\omega_m(\Delta t_1' + \Delta t_2')) \\
&\quad \times \boldsymbol{T}(-\Delta t_2)\boldsymbol{M}(\Delta x_2, -\omega_m \Delta t_1')\boldsymbol{T}(-\Delta t_1)\boldsymbol{M}(\Delta x_1, 0) \\
&= \boldsymbol{T}(-\Delta t_n)\boldsymbol{M}(\Delta x_n, -\omega_m(\Delta t_1' + \Delta t_2' + \cdots + \Delta t_n')) \times \cdots \\
&\quad \times \boldsymbol{T}(-\Delta t_3)\boldsymbol{M}(\Delta x_3, -\omega_m(\Delta t_1' + \Delta t_2')) \\
&\quad \times \boldsymbol{T}(-\Delta t_2)\boldsymbol{T}(-\Delta t_1)\boldsymbol{M}(\Delta x_2, 0)\boldsymbol{M}(\Delta x_1, 0) \\
&= \boldsymbol{T}(-\Delta t_n)\boldsymbol{M}(\Delta x_n, -\omega_m(\Delta t_1' + \Delta t_2' + \cdots + \Delta t_n')) \times \cdots \\
&\quad \times \boldsymbol{T}(-\Delta t_3)\boldsymbol{M}(\Delta x_3, -\omega_m(\Delta t_1' + \Delta t_2')) \\
&\quad \times \boldsymbol{T}(-(\Delta t_1 + \Delta t_2))\boldsymbol{M}(\Delta x_2 + \Delta x_1, 0),
\end{aligned}
\tag{6.107}
$$

where (6.95) and $\boldsymbol{T}(t_1)\boldsymbol{T}(t_1) = \boldsymbol{T}(t_1 + t_2)$ are used.

Similarly, by assuming $\Delta t_2 = \Delta t_2'$, we get

$$
\begin{aligned}
\boldsymbol{M}(\Delta x_3, &-\omega_m(\Delta t_1' + \Delta t_2'))\boldsymbol{T}(-(\Delta t_1 + \Delta t_2)) \\
&= \boldsymbol{T}(-(\Delta t_1 + \Delta t_2))\boldsymbol{M}(\Delta x_3, -\omega_m(\Delta t_1' + \Delta t_2')),
\end{aligned}
\tag{6.108}
$$

and

$$
\begin{aligned}
\boldsymbol{G} &= \boldsymbol{T}(-\Delta t_n)\boldsymbol{M}(\Delta x_n, -\omega_m(\Delta t_1' + \Delta t_2' + \cdots + \Delta t_n')) \times \cdots \\
&\quad \times \boldsymbol{T}(-(\Delta t_1 + \Delta t_2 + \Delta t_3))\boldsymbol{M}(\Delta x_3 + \Delta x_2 + \Delta x_1, 0).
\end{aligned}
\tag{6.109}
$$

By repeating this derivation, we get

$$
\begin{aligned}
\boldsymbol{G} &= \boldsymbol{T}(-(\Delta t_1 + \Delta t_2 + \cdots + \Delta t_n)) \\
&\quad \times \boldsymbol{M}(\Delta x_n + \cdots + \Delta x_2 + \Delta x_1, 0),
\end{aligned}
\tag{6.110}
$$

and

$$
\boldsymbol{G} = \boldsymbol{T}(-\Delta t)\boldsymbol{M}(x),
\tag{6.111}
$$

where the total delay by Δt is defined by

$$
\Delta t = \Delta t_1 + \Delta t_2 + \cdots + \Delta t_n.
\tag{6.112}
$$

Thus, when the delay of the modulating signal propagation on the electrode in the device is equal to that of the lightwave propagation in the waveguide in the device, the cascaded phase modulation processes with small amplitude modulating signals $\Delta x_k \sin \omega_m t$ provide the same output as with the modulation by the signal whose amplitude is x.

If difference in the propagation delays or the signal attenuation within the optical modulation device cannot be ignored, the modulation process can be analyzed by dividing the device into small parts in the matrix format described earlier. The equal propagation delay between the optical signal and the modulated signal corresponds to the velocity matching condition in the high-speed optical modulator for efficient optical phase modulation described in Sect. 4.2.4. When the propagation speeds of the modulation signal and the optical signal are equal, the propagation delay difference becomes zero, and, as discussed earlier, the amount of optical phase change in each part is added without loss.

Here, the matrix M depends only on the phase of the modulating signal $\sin \omega_m t$, while the delay of T is relative to the entire sideband component, represented by a vector of $2n + 1$-dimensions. For the modulating signal, the propagation delay is associated with the phase delay, which is division of the propagation distance by the phase speed, whereas the group delay in the optical waveguide is required for the optical signal. Therefore, the phase speed of the modulation signal and the group speed of the optical signal should be used to consider the velocity matching condition [4].

In general, the dispersion effect in the optical waveguides with weakly guiding such as titanium diffused waveguides, and that in the electrodes with transverse electromagnetic wave modes such as CPW, are small enough to neglect the difference between the phase and group velocity for analysis of the phase modulation. However, the effect of phase delay shift due to the dispersion in each element of T should be taken into account, when the dispersion effect in the waveguides is large.

6.2.5 Physical Interpretation of Sidebands Expanded by Bessel Functions

This section discusses the fundamental properties of each sideband component and its physical interpretation. For simplicity, we neglect the effect of the constant optical loss and phase delay in the device. If the amplitude of the sinusoidal wave signal applied to the modulator is z, the amplitude of the zeroth sideband component (carrier) is $J_0(z)$. As shown in (6.17), $J_0(z)$ is a decreasing function of z when $z \sim 0$ and is equal to unity at $z = 0$. On the other hand, for the other sideband components ($n \neq 0$), $J_n(0) = 0$, and $J_n(z)$ are increasing functions. As described, the behavior of the zeroth order sideband component is largely different from that of the other components in the range where z is not large ($|z| < 2$). $z = 0$ corresponds to an unmodulated signal, which has no sideband components. Thus, the amplitudes of the sidebands are zero, while that of the zeroth order component is unity. This is the physical interpretation of $J_0(0) = 1$ and $J_n(0) = 0$.

Increasing z results in the generation of the first and second sideband components. The energy of the sideband components is from the carrier component. Thus, the zeroth order sideband component decreases due to conversion from the carrier component to the other order sideband components. This is the physical explanation that $J_0(z)$ is a decreasing function.

As shown in Table 6.1, $J_1(z)$, which is an increasing function with respect to z and has a maximum at $z = 1.84$ and oscillatory change where $z > 1.84$. This means that the conversion rate from the zeroth to the first order components increases with the amplitude of the modulation signal z. Along with that, the component that is further converted from the first order sideband component to the higher order component (with some components returning to the zeroth order component) increases. The above maximum of $J_1(z)$ can be interpreted as the result of the competition of the two effects, i.e., the conversion processes outgoing and incoming to and from the first order component.

$J_0(z)$ has a zero at $z = 2.40$, and all energy is converted to other sideband components, but $J_0(z)$ turns into an increase as the amplitude of the modulated signal increases. It can be understood that this shows the effect that the energy dissipated into the other sideband components is converted back to the zeroth component by the effect of the modulation. For large z, $J_n(z)$ changes vibrationally with damping, as shown in (6.18). As the amplitude of the modulation signal increases, the energy expands to the higher order sideband components, so that the energy of the individual sideband components decreases on average. The oscillatory feature reflects the balance between conversion from and to the other sideband components.

As discussed in Sect. 6.2.4, the same conclusion can be achieved by breaking down a modulator into multiple cascaded modulators or grouping them together. Figure 6.20 shows only the 0-th, +1-th, and −1-th order sideband components in the cascaded phase modulation consisting of two modulation processes. At the first modulation process, the ±1-th order sideband components are generated from the zeroth order component. At the second modulation process, the second order (+2) component is generated from the first order (+1) component, while a part of the first order component is converted from the zeroth order component in the result of the first modulation process. Other sideband components including −2-th order component are generated in the same way. Thus, repeated phase modulation processes dissipate optical energy into various sideband components, where a wide variety of modulators are based on optical phase modulation. On the other hand, it is generally difficult to concentrate energy on a specific frequency component without loss. As mentioned in Sect. 4.3.1, lightwaves can be splitted by Y-junctions, so that we can easily obtain optical signals distributed in space domain. However, it would be rather difficult to concentrate the distributed lightwaves into a particular location without loss. It is easy to spread spatially, or in other words, to dissipate, but various devices are necessary to concentrate spatially spread lightwave in reverse. Since the optical phase modulation by the sinusoidal wave has the role of diffusing the optical energy in the frequency axis, it can be said that there is an analogy between the optical phase modulation and the Y-junctions as optical power diffusers.

6.3 Sideband Generation by a Mach–Zehnder Modulator

This section describes sideband generation in an MZM consisting of two parallel phase modulators. As described in Sect. 6.1, an optical phase modulator, to which a sinusoidal signal is fed, generates sideband components that can be expressed by the Bessel functions. The output of the MZM can be expressed as a sum of the sidebands generated by the two phase modulators, where the intensity of the output largely depends on the phase differences between the sideband components. The

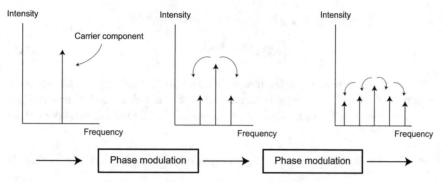

Fig. 6.20 Sideband generation by cascaded phase modulation processes

phase differences can be controlled by the phase of the modulation signal and the bias condition of the MZM, which is the optical phase difference between each phase modulator. This section offers mathematical expressions for the optical outputs from modulators consisting of two or more modulators connected in parallel, to discuss the optical output characteristics, including the effects of the intrinsic chirp parameter α_0, which represents the deviation from the ideal push–pull operation, and the extinction ratio, which represents the imbalance in the MZMs.

6.3.1 Phase Modulators Connected in Parallel

Consider the case where a sine wave signal is applied to two optical phase modulators connected in parallel that make up an MZM. To achieve ideal amplitude modulation, as discussed in Sect. 4.3.6, the two phase modulators must be well-balanced, where the sine waves are of equal amplitude with their signs reversed from each other. In actual modulators, it is necessary to take into account the amplitude and phase deviation generated inside the drive circuit or modulator.

Assuming that angular frequency of the modulation signal is ω_m and that A_1 and A_2 are the amplitudes of the induced optical phase shift by the phase modulation with the sine waves, whose phases are denoted by ϕ_1 and ϕ_2, the change of the optical phase at each phase modulator described in (4.34) can be written as

$$v_1(t) = A_1 \sin(\omega_m t + \phi_1) + B_1 \tag{6.113}$$

$$v_2(t) = A_2 \sin(\omega_m t + \phi_2) + B_2. \tag{6.114}$$

When the transmission coefficients of the two phase modulators are denoted by K_1 and K_2, the output R is given by

$$R = e^{i\omega_0 t} \left[K_1 e^{i\{A_1 \sin(\omega_m t + \phi_1) + B_1\}} \right.$$
$$\left. + K_2 e^{i\{A_2 \sin(\omega_m t + \phi_2) + B_2\}} \right]. \tag{6.115}$$

As described in Sect. 4.3.1, the transmission coefficients $K_1, K_2 = 1/2$, when the structure of the MZM is perfectly balanced, and the loss in the optical material and waveguide is zero. By using the matrix shown in (6.81), the optical output can be expressed by

$$R = e^{i\omega_0 t} \, e(\omega_m) \cdot \left[e^{iB_1} K_1 M(A_1, \phi_1) + e^{iB_2} K_2 M(A_2, \phi_2) \right] a_0, \tag{6.116}$$

where the optical input a_0 is assumed to be a monochromatic lightwave and is given by

$$a_0 \equiv \begin{bmatrix} \vdots \\ 0 \\ 0 \\ 1 \\ 0 \\ 0 \\ \vdots \end{bmatrix} \tag{6.117}$$

While the MZM consists of two parallel optical phase modulators, we can consider more complicated modulators which have more than two optical phase modulators as shown in Fig. 6.21. If the number of the optical phase modulators connected in parallel is N, the optical output is given by

$$R = e^{i\omega_0 t} \, e(\omega_m) \cdot \sum_{i=1}^{N} e^{iB_i} K_i M(A_i, \phi_i) a_0. \tag{6.118}$$

If the modulator has an ideal device structure without any loss where the splitting ratio in the balance is perfectly balanced, the amplitudes of the sideband components from each phase modulator would be $1/N$ at the optical output. Thus, the transmission coefficient of each phase modulator, K_i, is equal to $1/N$. In actual modulators with some imbalance, the difference from the ideal device is indicated by $\tilde{K}_i \equiv N K_i$. For prospective discussion, the output R can be expressed by

$$R = \frac{1}{N} e^{i\omega_0 t} \, e(\omega_m) \cdot \sum_{i=1}^{N} e^{iB_i} \tilde{K}_i M(A_i, \phi_i) a_0. \tag{6.119}$$

The case of $n = 2$ corresponds to the MZM consisting of two phase modulators, where $\tilde{K}_1 = 2K_1$ and $\tilde{K}_2 = 2K_2$. If the modulator has no loss and perfectly balanced structure, $\tilde{K}_1 = \tilde{K}_2 = 1$.

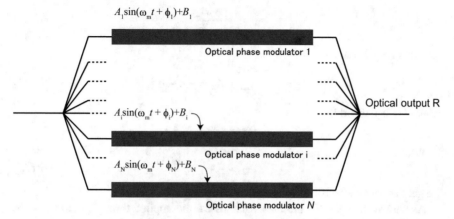

$A_1\sin(\omega_m t + \phi_1)+B_1$

Optical phase modulator 1

$A_i\sin(\omega_m t + \phi_i)+B_i$

Optical output R

Optical phase modulator i

$A_N\sin(\omega_m t + \phi_N)+B_N$

Optical phase modulator N

Fig. 6.21 Optical phase modulators connected in parallel

6.3.2 Mathematical Expressions of Mach–Zehnder Modulator Output

An optical output of an MZM consists of two optical signals generated by two optical phase modulators. The amplitudes of the sideband components in the optical output depend on the phase difference between the two optical signals at the Y-junction. It means that the spectrum can be changed by the bias condition, which is the optical path difference between the two optical phase modulators. The bias condition defined in Sect. 4.3.3 is given by

$$\Delta B \equiv \frac{B_1 - B_2}{2}, \tag{6.120}$$

where DC components of induced phase changes at the two optical phase modulators are denoted by B_1 and B_2 as shown in (6.115). The DC optical phase difference between the two optical phase modulators is described by $2\Delta B$.

While $2\Delta B = 0, \pm 2\pi, \pm 4\pi, \cdots$ corresponds to the full bias condition, $2\Delta B = \pm \pi, \pm 3\pi, \cdots$ corresponds to the null-bias condition. The quadrature bias condition is described by $2\Delta B = \pm \pi/2, \pm 3\pi/2, \cdots$. On the other hand, the amplitudes of RF components in the induced phases are described by A_1 and A_2.

$$B = \frac{B_1 + B_2}{2} \tag{6.121}$$

shows the average optical phase shift in the MZM output with respect to the optical input; however, this factor does not affect the output spectrum or the total system performance, as discussed in Sect. 4.5.

The intrinsic chirp parameter α_0 depends on the modulator structure and the configuration of the driving circuit as described in Sect. 4.3.6. From (4.49), we can derive the relation between the intrinsic chirp parameter α_0 and the amplitude of the RF components A_1, A_2 as follows:

$$\alpha_0 = \frac{A_1 + A_2}{A_1 - A_2}. \tag{6.122}$$

This can be rewritten as

$$A_1 = A + \alpha_A, \quad A_2 = -A + \alpha_A, \tag{6.123}$$

where α_A is defined by

$$\alpha_A \equiv A\alpha_0. \tag{6.124}$$

When

$$A_2 = -A_1, \tag{6.125}$$

we get the balanced push–pull operation where $\alpha_0 = 0$.

As shown in Fig. 4.47, the bias voltages can be applied through electrodes for the modulating signals that generate sideband components. In this case, the ratio between B_1 and B_2 (DC components in optical phase changes) is equal to the ratio between A_1 and A_2. The ratio that depends on the cross-section of the MZM cannot be controlled by the external driving circuit. On the other hand, when each phase modulator in the MZM has a separate electrode for dc bias, as shown in Fig. 4.48, the bias components B_1 and B_2 can be controlled independently. The average phase shift denoted by B does not affect the time domain profile or the frequency domain spectrum of the optical output, when the time domain fluctuation of B is slow enough. For example, if the spectral components of the fluctuation of B are confined in a low-frequency region whose frequency is lower than 1 kHz, the phase fluctuation of the optical output is dominated only by that of the laser source whose linewidth is larger than 1 kHz. Thus, it is not necessary to control B for optical modulators in most of the optical communication systems, while the spectrum depends on the bias, ΔB. In the following discussion, we assume that B_1 and B_2 can be set to arbitrary values to control the bias.

On the other hand, the ratio between the amplitudes of the high-frequency components, A_1 and A_2, is defined by the intrinsic chirp parameter α_0, which describes the imbalance of the phase modulation by the sinusoidal signal in the MZM. Here, we also consider the effect of the imbalance of the optical intensities in the two phase modulators, as described in Sect. 4.3.4. The optical output from the MZM can be expressed by

$$
\begin{aligned}
R = K\frac{e^{i\omega_0 t}}{2} &\left[e^{i\{A_1 \sin(\omega_m t + \phi_1) + B_1\}} \left(1 + \frac{\eta}{2}\right) \right. \\
&\left. + e^{i\{A_2 \sin(\omega_m t + \phi_2) + B_2\}} \left(1 - \frac{\eta}{2}\right) \right]
\end{aligned} \tag{6.126}
$$

$$
\begin{aligned}
= K\frac{e^{i\omega_0 t}}{2} &\left[e^{i\{(A+\alpha_A) \sin(\omega_m t + \phi_1) + B_1\}} \left(1 + \frac{\eta}{2}\right) \right. \\
&\left. + e^{i\{(-A+\alpha_A) \sin(\omega_m t + \phi_2) + B_2\}} \left(1 - \frac{\eta}{2}\right) \right],
\end{aligned} \tag{6.127}
$$

where the transmittances, K_1 and K_2, are defined by (4.54) and (4.55). η denotes the imbalance in the transmittance, due to fabrication error. As shown in (4.36), the overall loss in the modulator is described by K, which is equal to unity when the loss in the MZM is zero. The modulator with $\eta = 0$ has symmetric optical Y-junctions, i.e., a balanced interferometer, where $K_1 = K_2$. The balance in the modulation depth, i.e., in A_1 and A_2, is described by α_0. When the absolute values of η and α_0 are small, the MZM has high symmetry and provides almost ideal push–pull operation. The overall loss K does not affect the spectrum or the time domain profile of the output, so we assume that $K = 1$ in the following discussion, for simplicity.

By using the expansion by the Bessel functions described in (6.62) for the first and the second terms in (6.115), we get

$$
R = \frac{e^{i\omega_0 t}}{2} \sum_{n=-\infty}^{\infty} e^{in\omega_m t} \left[J_n(A_1) e^{in\phi_1 + iB_1} \left(1 + \frac{\eta}{2}\right) \right.
$$
$$
\left. + J_n(A_2) e^{in\phi_2 + iB_2} \left(1 - \frac{\eta}{2}\right) \right] \tag{6.128}
$$
$$
= \frac{e^{i\omega_0 t}}{2} \sum_{n=-\infty}^{\infty} e^{in\omega_m t} \left[J_n(A + \alpha_A) e^{in\phi_1 + iB_1} \left(1 + \frac{\eta}{2}\right) \right.
$$
$$
\left. + J_n(-A + \alpha_A) e^{in\phi_2 + iB_2} \left(1 - \frac{\eta}{2}\right) \right]. \tag{6.129}
$$

This is identical to elements of the matrix shown in (6.116). The n-th order sideband component power (the square of the amplitude), P_n can be expressed by

$$
P_n = E_n^2 \tag{6.130}
$$

$$
E_n = \frac{1}{2} \left| J_n(A + \alpha_A) e^{in\phi_1 + iB_1} \left(1 + \frac{\eta}{2}\right) \right.
$$
$$
\left. + J_n(-A + \alpha_A) e^{in\phi_2 + iB_2} \left(1 - \frac{\eta}{2}\right) \right|
$$
$$
= \frac{1}{2} \left| e^{iB + in\frac{\phi_1 + \phi_2}{2}} \left[J_n(A + \alpha_A) e^{i(n\frac{\phi}{2} + \Delta B)} \left(1 + \frac{\eta}{2}\right) \right.\right.
$$
$$
\left.\left. + J_n(-A + \alpha_A) e^{-i(n\frac{\phi}{2} + \Delta B)} \left(1 - \frac{\eta}{2}\right) \right] \right|
$$
$$
= \frac{1}{2} \left| J_n(A + \alpha_A) e^{i(n\frac{\phi}{2} + \Delta B)} \left(1 + \frac{\eta}{2}\right) \right.
$$
$$
\left. + J_n(-A + \alpha_A) e^{-i(n\frac{\phi}{2} + \Delta B)} \left(1 - \frac{\eta}{2}\right) \right|, \tag{6.131}
$$

where E_n is the absolute value of the amplitude of the n-th order sideband component. The phase difference between the two modulating signals ϕ defined by

$$
\phi \equiv \phi_1 - \phi_2, \tag{6.132}
$$

is called skew.

Here, we define,

$$\varphi_B \equiv 2\Delta B, \tag{6.133}$$

which is called bias, henceforth. ΔB corresponds to the average optical phase difference at each phase modulator.

Equation (6.129) can be rewritten by using ϕ and ϕ_B as follows:

$$R = \frac{e^{i\omega_0 t + iB_2}}{2} \sum_{n=-\infty}^{\infty} e^{in\omega_m t + in\phi_2} \left[J_n(A + \alpha_A)e^{in\phi + i\phi_B} \left(1 + \frac{\eta}{2}\right) \right.$$
$$\left. + J_n(-A + \alpha_A) \left(1 - \frac{\eta}{2}\right) \right], \tag{6.134}$$

where the absolute value of the sideband component is given by

$$E_n = \frac{1}{2} \left| J_n(A + \alpha_A)e^{in\phi + i\phi_B} \left(1 + \frac{\eta}{2}\right) \right.$$
$$\left. + J_n(-A + \alpha_A) \left(1 - \frac{\eta}{2}\right) \right|. \tag{6.135}$$

The relationship between the optical phase difference and the sideband amplitude can be clearly expressed by using ΔB, while ϕ_B and ϕ are also useful parameters offering concise expressions for the sideband components.

In addition, $\Phi = n\phi + \phi_B$ can provide another simple expression of the sideband components as follows:

$$E_n = \frac{1}{2} \left| J_n(A + \alpha_A)e^{i\Phi} \left(1 + \frac{\eta}{2}\right) + J_n(-A + \alpha_A) \left(1 - \frac{\eta}{2}\right) \right|$$
$$= \frac{1}{2} \left| J_n(A + \alpha_A) \left(1 + \frac{\eta}{2}\right) \cos\Phi + J_n(-A + \alpha_A) \left(1 - \frac{\eta}{2}\right) \right.$$
$$\left. + i J_n(A + \alpha_A) \left(1 + \frac{\eta}{2}\right) \sin\Phi \right|$$
$$= \frac{1}{2} \left[\left\{ J_n(A + \alpha_A) \left(1 + \frac{\eta}{2}\right) \right\}^2 + \left\{ J_n(-A + \alpha_A) \left(1 - \frac{\eta}{2}\right) \right\}^2 \right.$$
$$\left. + 2 J_n(A + \alpha_A) J_n(-A + \alpha_A) \left(1 - \frac{\eta^2}{4}\right) \cos\Phi \right]^{1/2}. \tag{6.136}$$

6.3.3 Balanced Mach–Zehnder Modulator

Here, we consider an ideal MZM with well-balanced optical loss in the waveguides ($\eta = 0$) and well-balanced push–pull operation ($\alpha_A = 0$). When the skew ϕ is zero and $A_1 = -A_2$, we can obtain the push–pull operation. In most of the MZMs using x-cut or z-cut LN substrate, we can assume $\phi = 0$, where $\phi_1 = \phi_2 = \tilde{\phi}$. When multiple sinusoidal signals are simultaneously applied on the MZM, the optical output spectrum depends on the phase of each sinusoidal signal, $\tilde{\phi}$. However, when a single tone sinusoidal signal is fed to a modulator, we can assume $\tilde{\phi} = 0$ without loss of generality. The output given by (6.129) can be rewritten as

$$R = \frac{e^{i\omega_0 t}}{2} \sum_{n=-\infty}^{\infty} e^{in(\omega_m t + \tilde{\phi})} \left[J_n(A)e^{iB_1} + J_n(-A)e^{iB_2} \right]$$

$$= \frac{e^{i\omega_0 t}}{2} e^{iB} \sum_{n=-\infty}^{\infty} e^{in(\omega_m t + \tilde{\phi})} J_n(A) \left[e^{i\Delta B} + (-1)^n e^{-i\Delta B} \right]. \qquad (6.137)$$

As with the phase modulation, each sideband component can be expressed by the Bessel function $J_n(A)$. The amplitude of the sidebands is given by

$$E_n = \frac{1}{2} \left| J_n(A)e^{i\Delta B} + (-1)^n J_n(A)e^{-i\Delta B} \right|$$

$$= \frac{1}{2} \left| J_n(A)e^{i\phi_B} + (-1)^n J_n(A) \right|. \qquad (6.138)$$

Thus, the amplitude of n-th order sideband component is proportional to

$$F_n(\Delta B) = e^{i\Delta B} + (-1)^n e^{-i\Delta B}, \qquad (6.139)$$

which describes the interference between the optical outputs from the two phase modulators embedded in the MZM. For the even order sidebands, where $(-1)^n = 1$, the factor shown in (6.139) is $2\cos\Delta B$. On the other hand, the factor for the odd order sidebands, where $(-1)^n = -1$, is $2i\sin\Delta B$. Thus, (6.137) can be rewritten as

$$R = e^{i\omega_0 t} e^{iB} \left[\cos\Delta B \left\{ \cdots + J_{-2}(A)e^{-2i(\omega_m t + \tilde{\phi})} + J_0(A) + J_2(A)e^{2i(\omega_m t + \tilde{\phi})} + \cdots \right\} \right.$$

$$+ i\sin\Delta B \left\{ \cdots + J_{-3}(A)e^{-3i(\omega_m t + \tilde{\phi})} + J_{-1}(A)e^{-i(\omega_m t + \tilde{\phi})} \right.$$

$$\left. \left. + J_1(A)e^{i(\omega_m t + \tilde{\phi})} + J_3(A)e^{3i(\omega_m t + \tilde{\phi})} + \cdots \right\} \right]. \qquad (6.140)$$

An even order sideband has a maximum when ΔB is equal to an integer multiple of π. The average phase difference between the optical signals from the two optical modulators, which is described by $2\Delta B (= \phi_B)$, is equal to an integer multiple of 2π. That means that the optical signals from the two phase modulators are combined together to interfere constructively, when the average optical path difference between the two phase modulators is equal to an integer multiple of the wavelength. On the other hand, the optical phase difference is an odd integer multiple of π. The odd order sideband components are combined with reversed phase to interfere destructively, when the optical path difference is zero.

Figure 6.22 shows the amplitude and phase of each sideband component. Here we consider the quadrature bias condition, as defined in Sect. 4.3.3, where the bias ΔB is equal to $\pi/4$; in other words, $\cos\Delta B = \sin\Delta B$. The phase of the modulating signal $\tilde{\phi}$ is assumed to be zero. The amplitude of the n-th order sideband is proportional to $J_n(A)$, so that the spectrum of the optical output without phase information, shown in Fig. 6.23, is identical to that of an optical output of a phase modulator. The

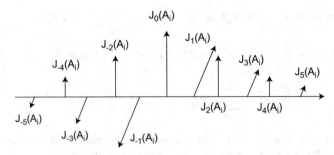

Fig. 6.22 Spectrum of the MZM optical output ($t = 0$)

intensities (the square of the amplitude) of the even order and odd order sidebands are proportional to $\cos^2 \Delta B$ and $\sin^2 \Delta B$, respectively.

Figure 6.24 shows the optical spectrum with a bias condition that is close to the full bias ($\Delta B \to 0$). The even order sidebands become maxima, while the odd order sidebands go to zero. When the average optical phase difference $2\Delta B$ is zero, the zeroth order sideband (carrier) whose optical frequency is equal to that of the optical input becomes a maximum by the constructive interference. The intensities of other even order sidebands go to maxima as well. On the other hand, the odd order sidebands are suppressed by destructive interference. For a bias condition that is close to the null bias ($\Delta B \to \pi/2$), the even order sidebands are suppressed, while the odd order sidebands go to maxima, as shown in Fig. 6.25.

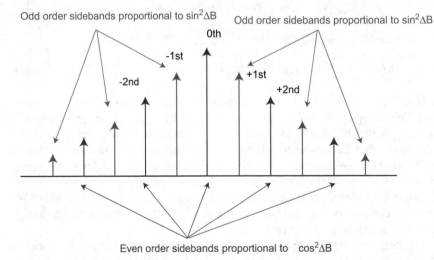

Fig. 6.23 MZM output spectrum with the quadrature bias condition ($\Delta B = \pi/4$)

The spectrum is symmetric with respect to the zeroth order component (y-axis), so that the output can be expressed by a product of a real function and $e^{i\omega_0 t} e^{iB}$ as follows:

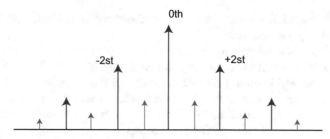

Fig. 6.24 MZM output spectrum with the full bias condition ($\Delta B \to 0$)

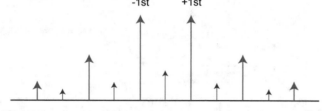

Fig. 6.25 MZM output spectrum with the null-bias condition ($\Delta B \to \pi/2$)

$$R = e^{i\omega_0 t} e^{iB} \left[\cos \Delta B \left\{ J_0(A) + 2 \sum_{n=1}^{\infty} J_{2n}(A) \cos[2n(\omega_m t + \tilde{\phi})] \right\} \right.$$
$$\left. - \sin \Delta B \cdot 2 \sum_{n=1}^{\infty} J_{2n-1}(A) \sin[(2n-1)(\omega_m t + \tilde{\phi})] \right]. \tag{6.141}$$

Thus, the absolute value of the optical output is given by

$$|R| = \cos \Delta B \left\{ J_0(A) + 2 \sum_{n=1}^{\infty} J_{2n}(A) \cos[2n(\omega_m t + \tilde{\phi})] \right\}$$
$$- 2 \sin \Delta B \sum_{n=1}^{\infty} J_{2n-1}(A) \sin[(2n-1)(\omega_m t + \tilde{\phi})]. \tag{6.142}$$

6.3.4 Basic Operation of Mach–Zehnder Modulator

In MZMs, the intensity of the optical output can be controlled by using optical interference between two optical signals from the two phase modulators. The optical phase difference between the two signals can be controlled through voltages applied to the modulators. As described in Sect. 4.3.1, the principle of the MZM can be explained by the interference between the two optical signals with the phase difference of ϕ_B ($= 2\Delta B$). When ϕ_B is equal to zero, the output intensity becomes a maximum due to constructive interference. When $\phi_B = \pi$, the optical path difference is equal to an integer multiple of a half wavelength, so that the output intensity goes to zero due to destructive interference. If the change of the optical phase difference ϕ_B is slow enough to be considered as quasi-static, the modulation does not have any significant impact on the optical spectrum.

Here, we consider modulation by a high-frequency signal that can be described by a sinusoidal waveform with DC bias ΔB $(= \phi_B/2)$. The ratio between different order sidebands within even order sidebands or odd order sidebands can be given by the Bessel function, as with the phase modulation. The even order sidebands are proportional to $\sin \Delta B$, while the odd order sidebands are proportional to $\cos \Delta B$. Thus, the square sum of these coefficient is constant.

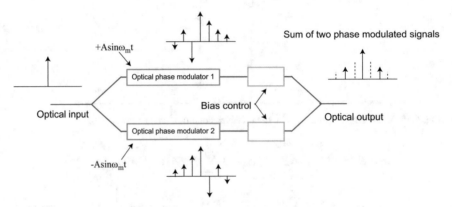

Fig. 6.26 Sideband components inside the MZM $(\Delta B = 0)$

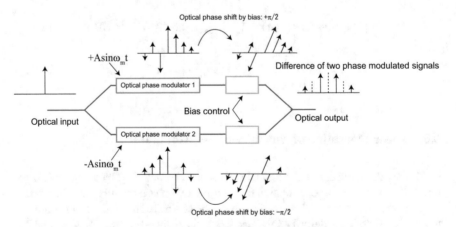

Fig. 6.27 Sideband components inside the MZM $(\Delta B = \pi/2)$

Sideband components at two phase modulators and Y-junctions with the full bias condition $(\Delta B = 0)$ are shown in Fig. 6.26, where we consider phase modulators with separate electrodes for the RF signal and the DC bias, to explain the sideband generation by the sinusoidal signal and the optical phase shift by the DC bias, individually. For a phase modulator with one electrode, the sinusoidal signal and the DC bias can be combined by an external circuit, where the operation principle is same as in the modulator with the two separate electrodes.

In the output of the phase modulator 1, the phases of all the USB components are equal to zero (upward on this page). On the other hand, the phases of the even order LSB components are 180-degree (downward on this page), while the even order USB components are aligned in 0-degree (upward on this page). In the output of the phase modulator 2, all the LSB components are in the same phase, 0-degree, while the even and odd order USB components have opposite polarities. The phase of the even order USB is 180-degree.

When the DC bias is set to the full bias ($\Delta B = 0$), the two optical signals from the phase modulators are summed in phase to generate the optical output. The odd order sidebands are suppressed by destructive interference, where the phase difference between the sideband components in the same order from the phase modulator 1 and the phase modulator 2 is 180-degree (π). Thus, the output consists only of the even order sideband components that are enhanced by constructive interference.

In the case of the null-bias condition ($\Delta B = \pi/2$), the phases of the sideband components are shifted by the DC bias voltage as shown in Fig. 6.27. The phase of the output from the phase modulator 1 is rotated by 90-degree ($+\pi/2$), while that from the phase modulator 2 is rotated by -90-degree ($-\pi/2$). Thus, the optical phase of the output from the phase modulator 2 is rotated by 180-degree with respect to the phase of the zeroth order (carrier) component in the output from the phase modulator 1. The MZM output, generated by combining these two optical signals at the Y-junction, consists of the difference of the two optical outputs from the phase modulators 1 and 2. Thus, the even order sidebands are suppressed, while the odd order sidebands are in the MZM output. Details of the MZM output are described in Sect. 7.1.

6.3.5 Skew in Modulating Signals

As shown in (6.131), the zeroth order component does not depend on the skew. For the other sideband components, $P_n \neq P_{-n}$ in general, when the skew is not equal to zero. Thus, the spectrum is not symmetric with respect to the zeroth order component.

For simplicity, we consider an ideal balanced MZM, where $\alpha_0 = \eta = 0$. The output R can be expressed by

$$R = \frac{e^{i\omega_0 t + iB_2}}{2} \sum_{n=-\infty}^{\infty} J_n(A) e^{in\omega_m t + in\phi_2} \left[e^{i(\phi_B + n\phi)} + (-1)^n \right]. \tag{6.143}$$

Thus, P_n is given by

$$P_n = \frac{J_n^2(A)}{2} \left[(-1)^n \cos \Phi + 1 \right]. \tag{6.144}$$

For the full bias condition ($\phi_B = 0$),

$$P_n = \frac{J_n^2(A)}{2} \left[(-1)^n \cos n\phi + 1 \right]. \tag{6.145}$$

For the null-bias condition ($\phi_B = \pi$),

$$P_n = \frac{J_n^2(A)}{2} \left[-(-1)^n \cos n\phi + 1 \right]. \tag{6.146}$$

Because $\cos(x)$ is an even function, $P_n = P_{-n}$ for arbitrary skew. The spectrum is symmetric with respect to the carrier component, when the bias ϕ_B is equal to 0 or π. We can confirm that the DC bias is set to the full- or null-bias condition, by changing the skew. If the spectrum is symmetric for arbitrary skew, ϕ, the bias ϕ_B is equal to 0 or π. This feature can be used for precise bias control.

If the bias condition is not set to the full or null bias, we can suppress the n-th order USB or LSB component, i.e., $P_n = 0, P_{-n} \neq 0$, by a particular combination of ϕ and ϕ_B. This is called single sideband (SSB) modulation. Details are described in Sect. 7.2.

Problems

6.1 Expand the Bessel functions up to the second order: $J_0(z)$, $J_1(z)$, and $J_2(z)$ into Taylor series around $z = 0$ up to the fourth order.

6.2 Prove (6.68), which describes the energy conservation law.

6.3 Derive (6.69).

6.4 As shown in (6.69) and (6.70), the power difference between the exact expression and the approximation describes the accuracy in the numerical calculation. Calculate the sum of the sideband component:

$$P_{t4}^{(4)} = \{J_0(A)\}^2 + 2 \sum_{n=1}^{4} \{J_n(A)\}^2,$$

by using Bessel functions expressed by fourth degree polynomials of the optical phase shift A, and discuss numerical calculation errors in the approximations.

6.5 Derive (6.144).

References

1. T. Kawanishi, T. Sakamoto, M. Izutsu, High-speed control of lightwave amplitude, phase, and frequency by use of electrooptic effect. IEEE J. Select. Top. Quant. Electron. **13**(1), 79–91 (2007)
2. F. Mitschke, *Bessel Functions* (Springer, Berlin, Heidelberg, 2010), pp. 265–267
3. E. Grosswald, *Bessel Polynomials and Bessel Functions: Differential Equations and Their Solutions* (Springer, Berlin, Heidelberg, 1978), pp. 4–17
4. A. Chen, E.J. Murphy(eds.), *Broadband Optical Modulators* (CRC Press, Boca Raton, 2012)

Chapter 7
Double Sideband and Single Sideband Modulation

The spectrum of the output from an MZM depends on device configuration, skew, and bias condition. This chapter provides modulator configuration for various applications including signal processing. Modulation schemes can be divided into double sideband (DSB) and single sideband (SSB). DSB signals have symmetric spectra with respect to the carrier, while the intensities of USB and LSB of the same order are not equal in the SSB signals. Here, modulator configurations for various applications including signal processing are described by using the mathematical expressions shown in Chap. 6. The sideband components in the DSB and SSB signals can be expressed by the Bessel function.

In the DSB modulation, there is no phase difference between the modulating signals on the two phase modulators, i. e., the skew is zero. USB and LSB components are balanced where the intensities of the USB and LSB components in the same order are equal. On the other hand, in the SSB modulation, the USB or LSB components can be suppressed largely by controlling the skew and the bias.

In this chapter, we focus on the carrier (the 0-th order component) and the first order USB and LSB, which are commonly used as desired components in actual systems. The intensities of the higher order sidebands are much smaller than the first order sidebands. However, the higher order components should be suppressed largely for particular applications which require large spurious suppression ratio.

7.1 Double Sideband Modulation and Bias Condition

The spectrum of the DSB signal largely depends on the bias condition. The intensities of the sidebands in the same order are balanced, where the skew ϕ is zero. A DSB signal whose 0-th order component (carrier) is close to zero is called a double sideband suppressed carrier (DSB-SC) signal. DSB modulation with the carrier component is simply denoted by DSB, but it can be called double sideband with carrier (DSB-C) to distinguish from DSB-SC. This section explains the three bias conditions: quadrature bias, null bias, and full bias, which are commonly used in

© Springer Nature Switzerland AG 2022

T. Kawanishi, *Electro-optic Modulation for Photonic Networks*, Textbooks in Telecommunication Engineering, https://doi.org/10.1007/978-3-030-86720-1_7

actual systems. The bias of the MZM, corresponding to the average optical phase difference between the two arms, can be expressed by $\phi_B(= 2\Delta B)$, where $\phi_B = \pi/2$ for the quadrature bias, $\phi_B = \pi$ for the null bias and $\phi_B = 0$ for the full bias.

7.1.1 Quadrature Bias

Here, we consider the MZM output with the quadrature bias condition, where the two optical signals from the two optical phase modulators embedded in the MZM are combined with $90°$ phase difference. By substituting $\Delta B = \pi/4$ $(\phi_B = \pi/2)$ into (6.129), we get

$$R = \frac{e^{i\omega_0 t - i\frac{\pi}{4}}}{2} \sum_{n=-\infty}^{\infty} e^{in\omega_m t + in\phi_2} \left[iJ_n(A + \alpha_A)e^{in\phi} \left(1 + \frac{\eta}{2}\right) \right.$$
$$\left. + J_n(-A + \alpha_A) \left(1 - \frac{\eta}{2}\right) \right], \tag{7.1}$$

where we assume $B_2 = -B_1 = -\pi/4$ $(B = 0)$, without loss of generality. The spectrum does not depend on the average phase shift B which affects the absolute phase of the output. The absolute value of the n-th order sideband, E_{nQ}, can be expressed by

$$E_{nQ} = \frac{1}{2} \left| iJ_n(A + \alpha_A)e^{in\phi} \left(1 + \frac{\eta}{2}\right) + J_n(-A + \alpha_A) \left(1 - \frac{\eta}{2}\right) \right|. \tag{7.2}$$

When the skew ϕ is equal to zero, E_{nQ} can be written by

$$E_{nQ} = \frac{1}{2} \left[J_n^2(A + \alpha_A) \left(1 + \frac{\eta}{2}\right)^2 + J_n^2(-A + \alpha_A) \left(1 - \frac{\eta}{2}\right)^2 \right]^{1/2}, \tag{7.3}$$

where the output has balanced USB and LSB, i. e., $E_{nQ} = E_{-nQ}$. In general, $E_{0Q} \neq 0$, so that the modulation scheme can be categorized into the DSB-C modulation.

By assuming, $|A| \gg |\alpha_A|$, $\eta \sim 0$, we ignore any second or higher order terms of α_A and η, the sideband components can be expressed by

$$E_{nQ} \simeq \frac{1}{2} \left[\left\{ J_n(A) + \alpha_A J_n'(A) \right\}^2 (1 + \eta) \right.$$
$$\left. + \left\{ J_n(-A) + \alpha_A J_n'(-A) \right\}^2 (1 - \eta) \right]^{1/2}$$
$$\simeq \frac{1}{2} \left[J_n^2(A)(1 + \eta) + 2\alpha_A J_n(A)J_n'(A) \right.$$
$$\left. + J_n^2(-A)(1 - \eta) + 2\alpha_A J_n(-A)J_n'(-A) \right]^{1/2}, \tag{7.4}$$

where we used $J_n(A + \alpha_A) \sim J_n(A) + \alpha_A J_n'(A)$. $J_n'(A)$ is the derivative of $J_n(A)$. $J_n(A)$ should be an even or odd function, so that $J_n^2(A) = J_n^2(-A)$. A derivative of an odd function is an even function, and that of an even function is an odd function. On other words, one of $J_n(A)$ and $J'_n(A)$ is an even function, and another is an odd function. Thus, $J'_n(-A)J_n(-A) = -J'_n(A)J_n(A)$. By using these relations, the sideband components can be expressed by

$$E_{nQ} \simeq \frac{|J_n(A)|}{\sqrt{2}}, \tag{7.5}$$

where the shape of the spectrum is identical to that shown in Fig. 6.23, which is discussed in Sect. 6.3.3. This equation includes impact of deviation from the ideal push-pull operation described by α_A and η, where the second or higher order terms are neglected. The spectrum does not depend on α_A or η, which means that the output spectrum with the quadrature bias condition is robust against deviation from the ideal push-pull operation.

The spectral shape of the MZM output is identical to that of the phase modulated signal, where the amplitude of the n-th order sideband component is proportional to $J_n(A)$. However, the amplitude of the MZM output is $1/\sqrt{2}$ of that of the phase modulated signal, because a half of the optical energy is dissipated at the Y-junction where the two optical signals are combined.

Here, we consider the carrier and the first order USB and LSB components by comparing with phase modulated signals. As discussed above, the impact of α_A and η is negligible, so shat, we assume α_A and η equal zero.

As described in (6.62), the phase modulated signal can be expressed by

$$\begin{aligned} R &\simeq \mathrm{e}^{i\omega_0 t} \left[J_{-1}(A)\mathrm{e}^{-i\omega_m t} + J_0(A) + J_1(A)\mathrm{e}^{i\omega_m t} \right] \\ &= \mathrm{e}^{i\omega_0 t} \left[-J_1(A)\mathrm{e}^{-i\omega_m t} + J_0(A) + J_1(A)\mathrm{e}^{i\omega_m t} \right] \\ &= \mathrm{e}^{i\omega_0 t} \left[J_0(A) + 2i \sin \omega_m t J_1(A) \right], \end{aligned} \tag{7.6}$$

where the modulating signal is $A \sin \omega_m t$.

When the carrier and the first order USB are in phase, ($t = 0$), the first order LSB is opposite sign to them. as shown in Fig. 7.1.

From (7.6), we can obtain a simple expression for the power of the optical output ($|R|^2$), as follows,

$$\begin{aligned} |R|^2 &= \left[-J_1(A)\mathrm{e}^{-i\omega_m t} + J_0(A) + J_1(A)\mathrm{e}^{i\omega_m t} \right] \\ &\quad \times \left[-J_1(A)\mathrm{e}^{i\omega_m t} + J_0(A) + J_1(A)\mathrm{e}^{-i\omega_m t}. \right] \end{aligned} \tag{7.7}$$

The total optical power consists of the square of each sideband component and of cross terms between different order sidebands. Considering the case of small signal modulation ($|A| \ll 1$), we ignore terms second order or higher, such as $J_1^2(A)$. The total optical power is given by

$$
\begin{aligned}
|R|^2 &= J_0^2(A) + 2J_1^2(A) + [J_1(A)J_0(A) - J_0(A)J_1(A)]e^{-i\omega_m t} \\
&\quad + [J_1(A)J_0(A) - J_0(A)J_1(A)]e^{i\omega_m t} - J_1^2(A)e^{-2i\omega_m t} - J_1^2(A)e^{2i\omega_m t} \\
&\sim J_0^2(A) + [J_1(A)J_0(A) - J_0(A)J_1(A)]e^{-i\omega_m t} + [J_1(A)J_0(A) - J_0(A)J_1(A)]e^{i\omega_m t}.
\end{aligned}
\tag{7.8}
$$

The coefficient for $e^{i\omega_m t}$ and $e^{-i\omega_m t}$, $[J_1(A)J_0(A) - J_0(A)J_1(A)]$, is obviously equal to zero. Because $J_0^2(A) \simeq 1$, so that

$$
|R|^2 = 1,
\tag{7.9}
$$

which is constant and does not depend on the modulating signal, $A \sin \omega_m t$. This reflects that optical modulation does not affect the optical power. Here, we used the simplest approximation, $J_0(A) = 1$.

By using (6.13) and (6.17) for

$$
J_0^2(A) + 2J_1^2(A),
\tag{7.10}
$$

we get

$$
\begin{aligned}
J_0^2(A) + 2J_1^2(A) &\simeq \left(1 - \frac{A^2}{4}\right)^2 + \frac{A^2}{2} \\
&\simeq 1 - 2 \times \frac{A^2}{4} + \frac{A^2}{2} = 1.
\end{aligned}
\tag{7.11}
$$

This implies that the intensity of the carrier component is a constant, even if we include high order terms. The coefficients for $e^{i\omega_m t}$ and $e^{-i\omega_m t}$ consist of the product of the carrier and the first order USB, $J_1(A)J_0(A)$, and that of the carrier and the first order LSB, $-J_1(A)J_0(A)$. These two terms, which have the same absolute value and the opposite signs, cancel each other as shown in Fig. 7.1.

By using (7.6), the phase modulated signal can be expressed on a phasor diagram in time domain, as shown in Fig. 7.2, where the signal can be described by a sum of a constant vector, corresponding to the carrier component, and a small vibrating component. The initial phase is set to the phase of the carrier component, where the vector of the carrier is on the real axis. The sum is on a circle with this first order approximation, so that the intensity of the optical power would be constant. If we include higher order terms, the vector of the phase modulated signal would move precisely on the circle.

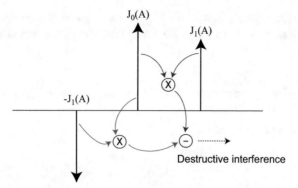

Fig. 7.1 Sidebands of a phase modulated signal

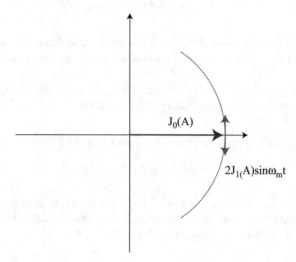

Fig. 7.2 Phasor of a phase modulated signal

By assuming α_A and η are zero in (7.1), the MZM output can be expressed by

$$
\begin{aligned}
R &= \frac{e^{i\omega_0 t - i\frac{\pi}{4}}}{2} \left[(i-1)J_{-1}(A)e^{-i\omega_m t} + (i+1)J_0(A) + (i-1)J_1(A)e^{i\omega_m t} \right] \\
&= \frac{e^{i\omega_0 t - i\frac{\pi}{4}}}{\sqrt{2}} \left[\frac{-i+1}{\sqrt{2}} J_1(A)e^{-i\omega_m t} + \frac{i+1}{\sqrt{2}} J_0(A) + \frac{i-1}{\sqrt{2}} J_1(A)e^{i\omega_m t} \right] \\
&= \frac{e^{i\omega_0 t - i\frac{\pi}{4}}}{\sqrt{2}} \left[J_1(A)e^{-i\omega_m t - i\frac{\pi}{4}} + J_0(A)e^{i\frac{\pi}{4}} + J_1(A)e^{i\omega_m t + i\frac{3\pi}{4}} \right] \\
&= \frac{e^{i\omega_0 t}}{\sqrt{2}} \left[J_1(A)e^{-i\omega_m t - i\frac{\pi}{2}} + J_0(A) + J_1(A)e^{i\omega_m t + i\frac{\pi}{2}} \right].
\end{aligned}
\tag{7.12}
$$

By using a time domain axis, $t' = t + \frac{\pi}{2\omega_m}$, where the difference between t and t' is equal to a quarter of the period of the modulating signal, the optical output can be expressed by

$$R = \frac{e^{i\omega_0 t' - i\frac{\pi\omega_0}{2\omega_m}}}{\sqrt{2}} \left[J_1(A)e^{-i\omega_m t'} + J_0(A) + J_1(A)e^{i\omega_m t'} \right]. \tag{7.13}$$

This equation can be rewritten as

$$R = \frac{e^{i\omega_0 t' - i\frac{\pi\omega_0}{2\omega_m}}}{\sqrt{2}} \left[J_0(A) + 2\cos\omega_m t' J_1(A) \right]$$

$$= \frac{e^{i\omega_0 t' - i\frac{\pi\omega_0}{2\omega_m}}}{\sqrt{2}} \left[J_0(A) - 2\sin\omega_m t J_1(A) \right]. \tag{7.14}$$

As shown in Fig. 7.3, the carrier, the first order USB and the first order LSB are in phase, when the phase difference between the carrier and the first order USB is zero, $(t' = 0)$.

Although the optical power, $|R|^2$ can be derived directly from (7.14), we express each sideband component by expanding (7.13), as follows:

$$|R|^2 = \frac{1}{2} \left[J_1(A)e^{-i\omega_m t'} + J_0(A) + J_1(A)e^{i\omega_m t'} \right]$$

$$\times \left[J_1(A)e^{i\omega_m t'} + J_0(A) + J_1(A)e^{-i\omega_m t'} \right], \tag{7.15}$$

which describes contribution of each sideband to the total optical power. As with the optical phase modulation, we ignore the second or higher order sideband components, such as $J_1^2(A)$, the total optical power can be expressed by

$$|R|^2 = \frac{1}{2} \left[J_0^2(A) + 2J_1^2(A) + 2J_0(A)J_1(A)e^{-i\omega_m t'} + 2J_0(A)J_1(A)e^{i\omega_m t'} \right.$$

$$\left. + J_1^2(A)e^{-2i\omega_m t'} + J_1^2(A)e^{2i\omega_m t'} \right]$$

$$= \frac{1}{2} \left[J_0^2(A) + 2J_1^2(A)(\cos 2\omega_m t' + 1) + 4J_0(A)J_1(A)\cos\omega_m t' \right]$$

$$= \frac{1}{2} \left[J_0^2(A) + 4J_1^2(A)\cos^2\omega_m t' + 4J_0(A)J_1(A)\cos\omega_m t' \right]. \tag{7.16}$$

By neglecting any second or higher order terms of A, the optical power can be approximated by

$$|R|^2 \sim \frac{1}{2} \left[1 + 4A\cos\omega_m t' \right]$$

$$= \frac{1}{2} \left[1 - 4A\sin\omega_m t' \right]. \tag{7.17}$$

The coefficients of $e^{i\omega_m t'}$ and $e^{-i\omega_m t'}$ consist of the product of the carrier and the first order USB, $J_1(A)J_0(A)$, and that of the carrier and the first order LSB, $-J_1(A)J_0(A)$, where these two terms have the same absolute value and the same sign. As shown in Fig. 7.3, these two terms interfere constructively, so that the optical output amplitude varies proportionally to the modulating signal, $A\sin\omega_m t$, which shows the basic principle of the intensity modulation by the MZM. The absolute value of the amplitude, $|R|$, is given by

$$|R| \sim \frac{1}{\sqrt{2}}\left[1 + 2A\cos\omega_m t'\right]$$
$$= \frac{1}{\sqrt{2}}\left[1 - 2A\sin\omega_m t\right]. \tag{7.18}$$

By using (7.14), the modulated signal can be described on a phasor diagram in time domain as shown in Fig. 7.4, where the modulated signal consists of a constant vector, corresponding to the carrier and a vibrating component, which means that the intensity of the output is modulated. These two vectors are parallel to each other, while there is no components perpendicular to the carrier. Thus, there is no phase change with the amplitude modulation.

When the bias ϕ_B is equal to $-\pi/2$, the output is

$$|R| \sim \frac{1}{\sqrt{2}}\left[1 + 2A\sin\omega_m t\right], \tag{7.19}$$

where the polarity of the vibrating component is flipped.

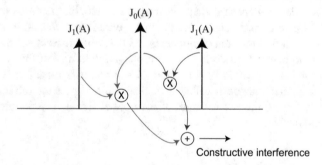

Fig. 7.3 Sideband of an MZM output with quadrature bias condition

As shown in Figs. 7.1 and 7.3, the phase modulated signal and the MZM output have the same spectral shape. These two signals are identical in the power spectral profile, however, they have difference in phase relations between the sideband components. In the intensity modulated signal by the MZM with the quadrature bias condition, the three components: the carrier, the first order USB and the first order LSB, are in phase at a particular time ($t' = 0$). On the other hand, in the phase

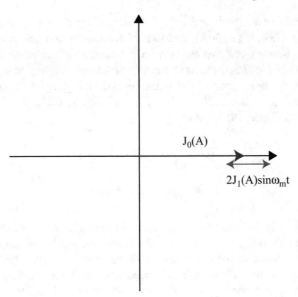

Fig. 7.4 Phasor of an intensity modulated signal

modulated signal, the polarity of the first order LSB is opposite to that of the carrier and the first order USB. Thus, the vibration of the intensity is suppressed due to the destructive interference between the product of the carrier and the USB and that of the carrier and the USB.

As shown here, even if the phase modulated and intensity modulated signals which have an identical spectral shape in frequency domain, the time domain profiles would have large difference. The temporal waveforms depend on the phase difference between the sideband components. As described in Sect. 4.3.5, the phase difference would be changed during signal propagation in optical fibers. Thus, the phase modulated signal would be converted into an intensity modulated signal and vice versa. This phenomenon is called fiber dispersion, where the refractive index in the fiber depends on the wavelength of the optical signal. The wavelength of a sideband component is

$$\lambda_n = \frac{2\pi c}{\omega_0 + n\omega_m}. \tag{7.20}$$

The angular frequency of the modulating signal ω_m is much less than ω_0. Thus, the phase shift due to the fiber dispersion would be negligible when the transmission distance is short enough.

However, in long distance transmission, the phase difference between the sideband components would be shifted due to the refractive index difference in the fiber. Even if the three components: the carrier, the first order USB and the first order LSB, are generated in phase by using an MZM with the quadrature bias condition, at a transmitter, the phase of the first order LSB would be flipped at a particular transmission distance, where the intensity modulation is diminished [1–3]. This effect depends on the frequency of the modulating signal and the transmission distance. An optical fiber link for data transmission would have dips in the frequency response where intensity modulation is converted into phase modulation due to the fiber dispersion effect. At the transmitter of the optical fiber link, wide frequency range signals should be applied to the modulator. This effect degrades waveforms over fibers significantly. The fiber dispersion can be compensated by using optical devices which has large dispersion whose polarity is opposite to that of the fiber. Recently, high-performance digital signal processing is commonly used for dispersion compensation for high-speed transmission systems with multilevel modulation formats [1, 4, 5].

7.1.2 Null Bias

By substituting $\Delta B = \pi/2 \quad (\phi_B = \pi)$ into (6.129), we get a mathematical expression for the output of an MZM with the null-bias condition, as follows,

$$R = \frac{e^{i\omega_0 t}}{2} \sum_{n=-\infty}^{\infty} e^{in\omega_m t + in\tilde{\phi}} \left[J_n(A + \alpha_A) \left(1 + \frac{\eta}{2}\right) \right.$$
$$\left. - J_n(-A + \alpha_A) \left(1 - \frac{\eta}{2}\right) \right], \tag{7.21}$$

where the skew $\phi = 0$ and $\phi_1 = \phi_2 = \tilde{\phi}$. For simplicity but without loss of generality, we assumed that $B_1 = 0 \quad (B = -\pi/2)$, where the polarity of the first term positive. This assumption affects only on the average phase.

The amplitudes of the sideband components for the null bias condition are

$$E_{nN} = \frac{1}{2} \left| J_n(A + \alpha_A) \left(1 + \frac{\eta}{2}\right) - (-1)^n J_n(A - \alpha_A) \left(1 - \frac{\eta}{2}\right) \right|. \tag{7.22}$$

Assuming $|\eta|, |\alpha_A/A| \ll 1$, we ignore the second or higher order terms of $|\eta|$ and $|\alpha_A/A|$. The sideband components can be expressed by

$$E_{nN} = \frac{1}{2} \left| [1 - (-1)^n] J_n(A) + [1 + (-1)^n] \left[\alpha_A J'_n(A) + \frac{\eta}{2} J_n(A) \right] \right|$$
$$= \begin{cases} |J_n(A)| & (n = 2m + 1) \\ |\alpha_A J'_n(A) + \frac{\eta}{2} J_n(A)| & (n = 2m) \end{cases}, \tag{7.23}$$

where m is an integer.

Amplitude of an odd order of E_{nN} can be described by the n-th order Bessel function, which does not depend on η and α_A. This equation includes the effect from the first order terms of η and α_A. Amplitude of an even order is sum of the effect of the interferometer imbalance (η) and that of the imbalance in the push-pull operation (α_A). When $\eta = 0$ and $\alpha_A = 0$, the amplitude of the even order sideband is equal to zero. Thus, we can obtain a DSB-SC signal where the even order sidebands including the carrier are suppressed. However, the output would contain residual even order sidebands due to finite α_A and η in actual MZM devices. Note that the spectrum of the MZM output with the null-bias condition is identical to the odd order sidebands of the phase modulated signal, as shown in (6.61), (6.62) and (7.24).

By using (6.140), the output from an ideal balanced MZM, where $\eta = 0$ and $\alpha_A = 0$, can be expressed, as follows:

$$R = \mathrm{i}e^{\mathrm{i}\omega_0 t}e^{\mathrm{i}B}\left[\cdots + J_{-3}(A)e^{-3\mathrm{i}(\omega_m t + \tilde{\phi})} + J_{-1}(A)e^{-\mathrm{i}(\omega_m t + \tilde{\phi})}\right.$$
$$\left. + J_1(A)e^{\mathrm{i}(\omega_m t + \tilde{\phi})} + J_3(A)e^{3\mathrm{i}(\omega_m t + \tilde{\phi})} + \cdots\right]. \tag{7.24}$$

This equation can be rewritten as

$$R = 2\mathrm{i}\sum_{n=1}^{\infty} J_{2n-1}(A)\sin[(2n-1)(\omega_m t + \tilde{\phi})]. \tag{7.25}$$

Using (6.13) and ignoring the third or higher order terms of A, the total output intensity is given by

$$|R|^2 \simeq 4J_1^2(A)\sin^2(\omega_m t + \tilde{\phi})$$
$$\simeq A^2\frac{1 - \cos 2(\omega_m t + \tilde{\phi})}{2}, \tag{7.26}$$

where the temporal profile varies with double the modulating-signal frequency.

For the even order sidebands, when

$$\eta = -\frac{2\alpha_A J_n'(A)}{J_n(A)}, \tag{7.27}$$

E_{nN} is equal to zero, where we assume that $J_n(A) \neq 0$. E_{nN} does not depend on η, when $J_n(A) = 0$. On the other hand, E_{nN} does not depend on α_A, when $J_n'(A) = 0$.

Here, we consider suppression of the carrier component, by using the condition described by (7.27), The amplitude of the carrier (the zeroth order sideband component) is given by

$$E_{0N} = \frac{1}{2}\left|J_0(A + \alpha_A)\left(1 + \frac{\eta}{2}\right) - J_0(A - \alpha_A)\left(1 - \frac{\eta}{2}\right)\right|, \tag{7.28}$$

which becomes zero with the condition of $\alpha_A = \eta = 0$. As described above, the carrier component can be suppressed perfectly by using the balance MZM with the push-pull operation.

On the other hand, when

$$\eta = \frac{2\left[-J_0(A + \alpha_A) + J_0(A - \alpha_A)\right]}{J_0(A + \alpha_A) + J_0(A - \alpha_A)}$$
$$\simeq -2\alpha_A\frac{J_0'(A)}{J_0(A)}, \tag{7.29}$$

the amplitude of the carrier component is zero, where the residual component generated by the imbalance in the interferometer and that by the imbalance in the push-pull operation cancel each other. However, the other even order sideband components would have finite residual amplitude. By substituting (7.29) into (7.22) for $n = 2$, we can obtain the residual second order sideband, as follows,

$$
\begin{aligned}
E_{2N} = \frac{1}{2} \Big| & J_2\left(A + \alpha_A\right) - J_2\left(A - \alpha_A\right) \\
& + \frac{-J_0\left(A + \alpha_A\right) + J_0\left(A - \alpha_A\right)}{J_0\left(A + \alpha_A\right) + J_0\left(A - \alpha_A\right)} \left\{ J_2\left(A + \alpha_A\right) + J_2\left(A - \alpha_A\right) \right\} \Big| \\
& \simeq \left| \alpha_A \left(J_2'(A) - J_0'(A) \frac{J_2(A)}{J_0(A)} \right) \right|.
\end{aligned}
\tag{7.30}
$$

By using the formulas of the derivatives of the Bessel functions,

$$
J_0'(A) = -J_1(A)
\tag{7.31}
$$

$$
J_2'(A) = J_1(A) - \frac{2J_2(A)}{A},
\tag{7.32}
$$

we get

$$
E_{2N}/E_{1N} = \alpha_0 \left[A - 2\frac{J_2(A)}{J_1(A)} + A\frac{J_2(A)}{J_0(A)} \right],
\tag{7.33}
$$

where the amplitude of the second order sideband is proportional to the intrinsic chirp parameter α_0. If the imbalance of the interferometer η is tunable, we can estimate α_0 from the second order sideband intensity [6–9].

Here, we consider an MZM with intensity trimmers embedded in the arms as shown in Fig. 7.5. The trimmers are also comprised of MZMs. The parameter η can be set to an arbitrary value by using the trimmers. The imbalance in the main MZM due to fabrication error can be compensated by tuning the bias voltages applied on the MZMs embedded in the interferometer arms. Thus, we can control η precisely, while the intrinsic chirp parameter α_0 can be controlled by using external circuits driving an MZM with two independent electrodes as shown in Fig. 7.5. In this configuration, in addition to the skew, $\phi_1 - \phi_2$, the parameters, α_0 and η, can be controlled independently. As described in (7.29) and (7.41), undesired sideband components can be largely suppressed by tuning η and α_0, simultaneously.

DSB-SC modulation with large spurious suppression has been demonstrated by using the MZM shown in Fig. 7.5. A DSB-SC signal consisting of the even order USB and LSB components can be generated by using the null-bias condition with trimming the imbalance in the interferometer and the push-pull operation. If the amplitude of the modulating signal A is small enough, the third or higher even order sidebands would be negligible. Thus, the DSB-SC signal has two spectral components whose amplitude is well-balanced and is called a two-tone signal. The two-tone signal can be used as a reference signal for radio astronomy, precise optoelectronic measurement,

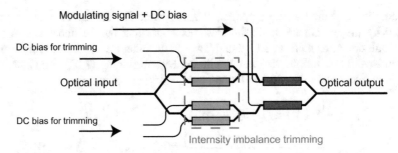

Fig. 7.5 MZM with intensity trimmers

etc., while the suppression of the spurious components (for example, the carrier and the second order sidebands), is an issue in conventional modulators [1, 6, 8].

Figure 7.6 shows an example of an optical spectrum of a DSB-SC signal, where the modulation frequency was 10.5 GHz. A two-tone signal whose frequency separation was 21 GHz, was generated by using the null-bias operation with high-extinction ratio ($\eta \sim 0$) and low-chirp parameter ($\alpha_0 \sim 0$), where the spurious suppression ratio was 46.8 dB. The extinction ratio can be estimated as 66 dB, from the suppression ratio of the carrier component. By substituting the ratio of the first and the second order sidebands into (7.33), we can deduce that $|\alpha_0| < 0.01$. The absolute value of the intrinsic chirp parameter was less than the measurement limit [6].

Figure 7.7 shows optical output spectra with varying α_0 through external driving circuits. The carrier component was minimized by trimming η. The intensity of the second order sideband is an increasing function of α_0. On the other hand, we can obtain the value of α_0 from the ratio between the first and the second order sideband components. As shown in Fig. 7.8, the intrinsic chirp parameter can be precisely estimated from the spectra. Various parameters including α_0 and η which describe modulator structures, operation conditions, etc., can be obtained from spectra of the optical output. Details will be described in Sect. 8.1.

7.1.3 Full Bias

By substituting $\Delta B = 0$ ($\phi_B = 0$), into (6.129), as with the previous section, the optical output can be expressed by

$$
R = \frac{e^{i\omega_0 t + iB_2}}{2} \sum_{n=-\infty}^{\infty} e^{in\omega_m t + i\phi_2} \left[J_n(A + \alpha_A)\left(1 + \frac{\eta}{2}\right) \right.
$$
$$
\left. + J_n(-A + \alpha_A)\left(1 - \frac{\eta}{2}\right) \right].
\tag{7.34}
$$

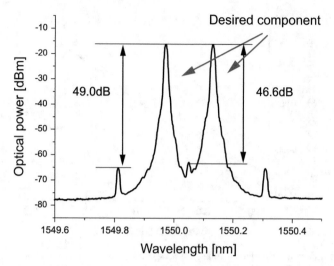

Fig. 7.6 Spectrum of a DSB-SC signal generated by a balanced MZM

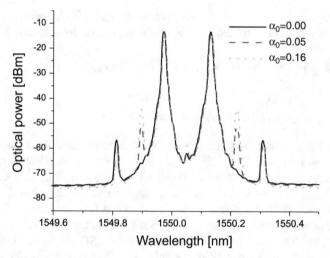

Fig. 7.7 Spectra of DSB-SC signals with varying the intrinsic chirp parameter

The sideband components with the full bias, E_{nF}, are given by

$$
\begin{aligned}
E_{nF} = \frac{1}{2} \bigg| &J_n(A + \alpha_A)\left(1 + \frac{\eta}{2}\right) \\
&+ (-1)^n J_n(A - \alpha_A)\left(1 - \frac{\eta}{2}\right) \bigg|.
\end{aligned}
\tag{7.35}
$$

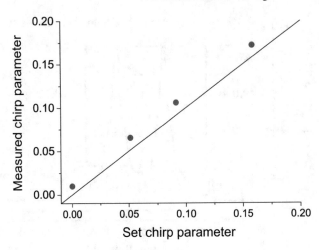

Fig. 7.8 Intrinsic chirp parameter estimated from spectra

By assuming that $|\eta|, |\alpha_A/A| \ll 1$, we neglect the second or higher order components of $|\eta|$ and $|\alpha_A/A|$. The amplitudes of the sidebands can be described by

$$E_{nF} = \frac{1}{2} \left| [1 + (-1)^n] J_n(A) + [1 - (-1)^n] \left[\alpha_A J_n'(A) + \frac{\eta}{2} J_n(A) \right] \right|$$

$$= \begin{cases} |J_n(A)| & (n = 2m) \\ |\alpha_A J_n'(A) + \frac{\eta}{2} J_n(A)| & (n = 2m + 1). \end{cases} \tag{7.36}$$

In contrast to the null bias, the absolute value of the even order sideband, E_{2mF} is equal to the n-th order Bessel function, while the even order component consists of the effects of η and α_A. When $\eta = 0$ and $\alpha_A = 0$, the amplitude of the odd order sideband, E_{2m+1F}, is zero. Thus, we can obtain the optical output consists of even order sidebands, while the odd order sidebands are largely suppressed. In addition to the carrier component, the major components in the output are the second order USB and LSB. The amplitudes of the USB and the LSB are equal to each other, so that the optical output is a DSB-C signal. As with the null bias, the odd order sidebands including the first order USB and LSB have residual components due to finite α_A and η.

In a similar way to the previous section, the output R of a balanced MZM can be given by

$$R = e^{i\omega_0 t} \left[\cdots + J_{-2}(A) e^{-2i(\omega_m t + \tilde{\phi})} + J_0(A) + J_2(A) e^{2i(\omega_m t + \tilde{\phi})} + \cdots \right], \tag{7.37}$$

where we assume that $B = 0$. This equation can be rewritten as

$$R = e^{i\omega_0 t} \left[J_0(A) + 2 \sum_{n=1}^{\infty} J_{2n}(A) \cos[2n(\omega_m t + \tilde{\phi})] \right]. \tag{7.38}$$

By using (6.14) and (6.17), we neglect the third or higher order terms of A, in other words, sidebands the third or higher order. The output power can be expressed by

$$|R|^2 \simeq J_0^2(A) + 4J_0(A)J_2(A)\cos[2n(\omega_m t + \tilde{\phi})]$$

$$\simeq 1 - \frac{A^2}{4} + \frac{A^2}{2}\cos[2n(\omega_m t + \tilde{\phi})]. \tag{7.39}$$

Similarly to the null bias, the temporal profile of the output varies with double the modulating-signal frequency. The average power of the output with the full bias is $1 - A^2/4$, while that with the null bias is $A^2/2$.

As with the suppression of the carrier component in the null bias, we consider the suppression of the first order sidebands. The amplitude of the first order sideband is

$$E_{1F} = \frac{1}{2}\left[J_1(A + \alpha_A)\left(1 + \frac{\eta}{2}\right) - J_1(A - \alpha_A)\left(1 - \frac{\eta}{2}\right)\right]. \tag{7.40}$$

Similarly to the carrier component (the zeroth order sideband) in the null bias, the first order sideband in the full bias becomes zero, when both α_A and η are equal to zero. In addition, when

$$\eta = \frac{2\left[-J_1(A + \alpha_A) + J_1(A - \alpha_A)\right]}{J_1(A + \alpha_A) + J_1(A - \alpha_A)}$$

$$\simeq -2\alpha_A\frac{J_1'(A)}{J_1(A)}, \tag{7.41}$$

the first order sidebands are suppressed, due to the destructive interference between the effect of the interferometer imbalance and that of the push-pull operation imbalance.

By substituting (7.41) into (7.35) for $n = 3$, we can obtain the residual third order sideband, as follows:

$$E_{3F} = \frac{1}{2}\Big[J_3(A + \alpha_A) - J_3(A - \alpha_A)$$

$$+ \frac{-J_1(A + \alpha_A) + J_1(A - \alpha_A)}{J_1(A + \alpha_A) + J_1(A - \alpha_A)}\{J_3(A + \alpha_A) + J_3(A - \alpha_A)\}\Big]$$

$$\simeq \alpha_A\left(J_3'(A) - J_1'(A)\frac{J_3(A)}{J_1(A)}\right). \tag{7.42}$$

The amplitude ratio of the third order sideband to the second order is proportional to the intrinsic chirp parameter, as described by

$$E_{3F}/E_{2F}$$

$$= \alpha_0\frac{AJ_1(A)J_2(A) - 2J_1(A)J_3(A) - AJ_0(A)J_3(A)}{J_1(A)J_2(A)}. \tag{7.43}$$

Thus, α_0 can be estimated from the spectra of the full bias, as with the null-bias condition.

7.2 Single Sideband Modulation and Optical Frequency Shift

USB and LSB generated by an MZM have amplitude imbalance, when the modulating signals applied to the two phase modulators embedded in the MZM has a finite skew, which is associated with the phase difference between the modulating signals. By tuning the skew properly, the USB or LSB in the modulator output can be largely suppressed. This modulation scheme is called single sideband (SSB) modulation, where the SSB signal has either the USB or LSB. Firstly, this section describes the SSB modulation using a simple MZM consisting of two phase modulators. The output has the carrier and one of the sidebands (USB and LSB). A dual-parallel MZM consisting of two MZMs can offer single sideband suppressed carrier (SSB-SC) modulation where the output has only one major spectral component consisting of one of the sidebands. The carrier component is largely suppressed in the two MZMs with the null-bias condition. When the effects of the high order sideband generation can be neglected, the SSB-SC modulation does not change the spectral shape of the input optical signal. The frequency difference between the input and output signals is equal to the frequency of a sinusoidal signal applied to the modulator. Thus, the DPMZM acts as an optical frequency shifter, where the center frequency of the optical output can be precisely controlled by the frequency of the modulating signal.

7.2.1 Single Sideband Modulation by a Simple Mach-Zehnder Modulator

Here, for simplicity, we consider a balanced MZM, where α_0 and $\eta = 0$. By using (6.144), we can obtain a condition which suppresses the first order USB whose power is given by

$$P_1 = \frac{J_1^2(A)}{2}[-\cos(\phi_B + \phi) + 1]. \qquad (7.44)$$

When

$$\phi_B + \phi = 0, \qquad (7.45)$$

P_1 becomes 0. On the other hand, the power of the first order LSB, P_{-1}, can be expressed by

$$
\begin{aligned}
P_{-1} &= \frac{J_1^2(A)}{2}[-\cos(\phi_B - \phi) + 1] \\
&= \frac{J_1^2(A)}{2}[-\cos 2\phi_B + 1] \\
&= J_1^2(A)\sin^2\phi_B.
\end{aligned}
\qquad (7.46)
$$

Thus, P_{-1} has a maximum at $\phi_B = \pm\pi/2$, when we assume that $-\pi < \phi_B \le \pi$.

Note that $P_{-1} = 0$ at $\phi_B = 0$ and π. $\phi_B = 0$ corresponds to the full bias condition where both the amplitudes of ± 1 order sideband components (the first order USB and LSB) are zero. When $\phi_B = \pi$, the skew $\phi = -\pi$. Thus, the phase difference between the modulating signals applied to the two phase modulators embedded in the MZM is π. However, the phase difference is defined with respect to the ideal push-pull operation, so that the two phase modulators generate identical signals. The optical phase difference of π corresponds to the null-bias condition, where the output is the difference of the two optical signals from the phase modulator. Thus, all the sideband components including the carrier, the first order USB and LSB, are diminished in the output. These operation conditions ($\phi_B = 0$ and π) suppress both the USB and LSB, so that we exclude them from our discussion below.

Similarly, when $\phi_B - \phi = 0$, the first order LSB is suppressed. P_1 is given by

$$P_1 = J_1^2(A) \sin^2 \phi_B. \tag{7.47}$$

When $\phi_B = \pm\pi/2$, the first order USB has maxima.

In these both cases ($\phi_B = \pm\phi$), the power of the carrier component can be expressed by

$$P_0 = J_1^2(A) \cos^2 \frac{\phi_B}{2}. \tag{7.48}$$

Because $\phi_B \ne \pi$, $P_0 \ne 0$. Thus, the output always has the carrier component.

For SSB modulation with the first order LSB, the phase differences in modulating and optical signals given by $(\phi_B, \phi) = (+\pi/2, -\pi/2)$ and $(\phi_B, \phi) = (-\pi/2, +\pi/2)$ maximize the desired component (the first order LSB), and suppress the undesired component (the first order USB). On the other hand, the phase differences of $(\phi_B, \phi) = (+\pi/2, +\pi/2)$ and $(\phi_B, \phi) = (-\pi/2, -\pi/2)$ maximize the first order USB with suppressing the first order LSB. These phase differences offer optimal SSB modulation conditions.

Here, we consider the quadrature bias condition of $\phi_B = \pi/2$. By using (6.143), the optical output with $\phi = \pi/2$ can be derived as follows,

$$
\begin{aligned}
R &= \frac{e^{i\omega_0 t - i\pi/4}}{2} \sum_{n=-\infty}^{\infty} J_n(A) e^{in\omega_m t - in\pi/4} \left[e^{i\pi/2} e^{in\pi/2} + (-1)^n \right] \\
&= \frac{e^{i\omega_0 t}}{2} \sum_{n=-\infty}^{\infty} J_n(A) e^{in\omega_m t + in\pi/2} \left[e^{i\pi/4} e^{-in\pi/4} + e^{-i\pi/4} e^{in\pi/4} \right] \\
&= e^{i\omega_0 t} \sum_{n=-\infty}^{\infty} e^{in\omega_m t} J_n(A) i^n \cos \frac{(n-1)\pi}{4}. \tag{7.49}
\end{aligned}
$$

By using up to the third order sideband components, it can be rewritten as follows:

$$R = e^{i\omega_0 t}\left[\cdots + i^{-3}e^{-3i\omega_m t}J_{-3}(A)\cos(-\pi)\right.$$

$$+ i^{-2}e^{-2i\omega_m t}J_{-2}(A)\cos\left(-\frac{3}{4}\pi\right) + i^{-1}e^{-i\omega_m t}J_{-1}(A)\cos\left(-\frac{1}{2}\pi\right)$$

$$+ J_0(A)\cos\left(-\frac{1}{4}\pi\right) + i e^{i\omega_m t}J_1(A)\cos(0)$$

$$+ i^2 e^{2i\omega_m t}J_2(A)\cos\left(\frac{1}{4}\pi\right) + i^3 e^{3i\omega_m t}J_3(A)\cos\left(\frac{1}{2}\pi\right)\cdots\left.\right]$$

$$= e^{i\omega_0 t}\left[\cdots + i e^{-3i\omega_m t}J_3(A) + \frac{1}{\sqrt{2}}e^{-2i\omega_m t}J_2(A) + \frac{1}{\sqrt{2}}J_0(A)\right.$$

$$\left. + i e^{i\omega_m t}J_1(A) - \frac{1}{\sqrt{2}}e^{2i\omega_m t}J_2(A) + \cdots\right]. \tag{7.50}$$

By neglecting any second order or higher sideband components, we get,

$$R \simeq e^{i\omega_0 t}\left[\frac{1}{\sqrt{2}}J_0(A) + i e^{i\omega_m t}J_1(A)\right]. \tag{7.51}$$

Here, we use a time axis defined by $t' = t + \frac{\pi}{2\omega_m}$. The optical output can be expressed by

$$R = e^{i\omega_0 t' - i\frac{\pi\omega_0}{2\omega_m}}\left[\cdots + e^{-3i\omega_m t'}J_3(A) - \frac{1}{\sqrt{2}}e^{-2i\omega_m t'}J_2(A) + \frac{1}{\sqrt{2}}J_0(A)\right.$$

$$\left. + e^{i\omega_m t'}J_1(A) + \frac{1}{\sqrt{2}}e^{2i\omega_m t'}J_2(A) + \cdots\right]$$

$$\simeq e^{i\omega_0 t' - i\frac{\pi\omega_0}{2\omega_m}}\left[\frac{1}{\sqrt{2}}J_0(A) + e^{i\omega_m t'}J_1(A)\right]. \tag{7.52}$$

Figure 7.9 shows the sideband components up to the third order in the frequency domain. The amplitudes of the desired upper sideband E_1 and that of the carrier component E_0 are equal to $J_1(A)$ and $J_0(A)/\sqrt{2}$, respectively, while the first order lower sideband is suppressed ($E_{-1} = 0$). Each amplitude of the second order USB and LSB is $E_{\pm 2} = J_2(A)/\sqrt{2}$. As shown in Fig. 7.9, the output is an SSB signal consisting of the zeroth order sideband (carrier component) and the first order USB, where the second order (USB and LSB), and the third order (LSB) sideband components are the major part of the undesired components.

The intensity of the output $|R|^2$ can be expressed approximately by

$$|R|^2 = \frac{J_0^2(A)}{2} + J_1^2(A) - \sqrt{2}J_0(A)J_1(A)\sin\omega_m t, \tag{7.53}$$

where the third term is proportional to the modulating signal. Thus, the envelop is similar to that of a DSB signal with the quadrature bias condition. The total bandwidth of the SSB signal is a half of the DSB signal, so that the SSB signal would be robust against the dispersion effect in lightwave propagation over fibers

[10, 11]. Even if the second order sideband components are taken into account, the product of the carrier and the second order USB, and that of the carrier and the second order LSB, have destructive interference not to affect the intensity profile.

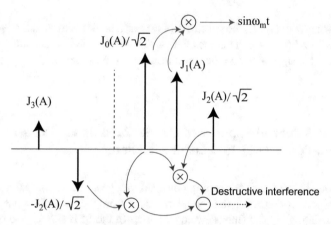

Fig. 7.9 SSB modulation by an MZM

Similarly, when $\phi = -\pi/2$,

$$R = e^{i\omega_0 t} \sum_{n=-\infty}^{\infty} e^{in\omega_m t} J_n(A) i^{-n} \cos \frac{(n+1)\pi}{4}$$

$$= e^{i\omega_0 t} \left[\cdots + i^3 e^{-3i\omega_m t} J_{-3}(A) \cos \left(-\frac{1}{2}\pi \right) \right.$$

$$+ i^2 e^{-2i\omega_m t} J_{-2}(A) \cos \left(-\frac{1}{4}\pi \right) + i e^{-i\omega_m t} J_{-1}(A) \cos (0)$$

$$+ J_0(A) \cos \left(\frac{1}{4}\pi \right) + i^{-1} e^{i\omega_m t} J_1(A) \cos \left(\frac{1}{2}\pi \right)$$

$$\left. + i^{-2} e^{2i\omega_m t} J_2(A) \cos \left(\frac{3}{4}\pi \right) + i^{-3} e^{3i\omega_m t} J_3(A) \cos (\pi) \cdots \right]$$

$$= e^{i\omega_0 t} \left[\cdots - \frac{1}{\sqrt{2}} e^{-2i\omega_m t} J_2(A) - i e^{-i\omega_m t} J_1(A) + \frac{1}{\sqrt{2}} J_0(A) \right.$$

$$\left. + \frac{1}{\sqrt{2}} e^{2i\omega_m t} J_2(A) - i e^{-3i\omega_m t} J_3(A) \cdots \right], \tag{7.54}$$

where $E_{-1} = J_1(A)$ and $E_1 = 0$. By neglecting the second order or higher sideband components, we get,

$$R \simeq e^{i\omega_0 t} \left[\frac{1}{\sqrt{2}} J_0(A) - i e^{-i\omega_m t} J_1(A) \right]. \tag{7.55}$$

By using $n' = -n$, the output R can be rewritten as

$$R = e^{i\omega_0 t} \sum_{n'=-\infty}^{\infty} e^{-in'\omega_m t} J_{-n'}(A) i^{n'} \cos \frac{(n'-1)\pi}{4}. \tag{7.56}$$

By comparing with (7.49), the spectrum of $\phi = \pi/2$ is identical to that of $\phi = -\pi/2$, when n' is rewritten by n. This means that the spectra of $\phi = \pi/2$ and $\phi = -\pi/2$ are the mirror images of each other with respect to the carrier frequency (the zeroth order).

7.2.2 Sigle-Sideband Suppressed Carrier Modulation Using a Dual-Parallel Mach-Zehnder Modulator

As described in the previous section, an MZM can generate an optical SSB signal, with $\pi/2$ phase differences in the optical bias condition (the quadrature bias) and also in the modulating signals (the skew equals 90°). In the SSB signal, one of the first order sidebands is suppressed and the other is maximized. The carrier component remains in the optical output. This section describes SSB-SC modulation by using a DPMZM consisting of two small MZMs embedded in a main MZM. The carrier is suppressed at the sub MZMs, while one of the USB and LSB would be suppressed at the Y-junction of the main MZM.

As shown in (7.24), an MZM with the null-bias condition suppresses the even order sidebands including the carrier component. Thus, the output has only the even order sidebands. One of the first order sidebands (USB or LSB) is suppressed by combining the two output from the two MZMs with the null-bias condition. A pair of sinusoidal signals which have $\pi/2$ phase difference are fed to the two MZMs embedded in a DPMZM, as modulating signals.[1]

The optical paths in the main MZM induces $\pi/2$ optical phase difference which can be controlled by a phase modulator located in the arms of the main MZM. Due to the destructive interference of one of the first order sidebands (USB or LSB), the optical output has one of the first order sidebands with suppression of the other sideband, similar to the SSB modulation with a simple MZM. In the SSB modulation with the DPMZM, the carrier component and the second order sidebands are suppressed in the two MZMs with the null-bias condition. Thus, the output has only one spectral component (the first order USB or LSB), when the third or higher order sidebands can be neglected. This is the basic principle of the SSB-SC modulation using the DPMZM.

The mathematical expressions of the SSB-SC signals generated by the DPMZM are similar to that of the SSB signals generated by the simple MZM. However, the even order sidebands are suppressed in the SSB-SC signals, while the SSB signals

---

[1] The skew ϕ is defined as the phase difference from the ideal push-pull operation condition, where the amplitudes of the induced optical phases on the two optical waveguides in an MZM is equal to π. Thus, the phase difference between the two modulating signals can be expressed by $\phi + \pi$.

generated by the simple MZM contain both the even and odd sidebands. Thus, the mathematical expressions of the SSB-SC signals can be obtained by eliminating terms associated with the even order sidebands from the mathematical expressions of the SSB signals: (7.50) and (7.54).

An SSB-SC signal with the bias $\phi_B = \pi/2$ and the skew $\phi = +\pi/2$ can be derived from (7.50), as follows:

$$R = e^{i\omega_0 t}\left[\cdots + ie^{-3i\omega_m t}J_3(A) + ie^{i\omega_m t}J_1(A)\cdots\right]$$
$$\simeq ie^{i(\omega_0+\omega_m)t}J_1(A). \tag{7.57}$$

Similarly, for the SSB-SC signal with $\phi = -\pi/2$, we can derive the mathematical expression from (7.54), as follows:

$$R = e^{i\omega_0 t}\left[\cdots - ie^{-i\omega_m t}J_1(A) - ie^{-3i\omega_m t}J_3(A)\cdots\right]$$
$$\simeq -ie^{i(\omega_0-\omega_m)t}J_1(A). \tag{7.58}$$

In either case, the output has only one spectral component whose frequency is shifted from that of the optical input. The frequency difference is equal to the frequency of the modulating signals (the sinusoidal signals applied to the two MZMs). As discussed in the Sect. 6.2.3, the optical phase modulation has the linearity for the optical input. Even if the output signal has a complicated waveform with various spectral components, the SSB-SC modulation does not change the temporal waveform and the spectral shape. Thus, the SSB-SC modulation can be used for optical frequency shift, where the output frequency can be controlled by the modulating signals.

The DPMZM, whose optical bias ϕ_B is set to $-\pi/2$, can offer the SSB-SC modulation, as with the case of the bias of $\pi/2$. When the phase differences, (ϕ_B, ϕ), are equal to $(\pi/2, \pi/2)$ or $(-\pi/2, -\pi/2)$, the output has the first order USB whose frequency is shifted upwards. On the other hand, when (ϕ_B, ϕ) are equal to $(\pi/2, -\pi/2)$ or $(-\pi/2, \pi/2)$, the DPMZM generates the first order LSB whose frequency is shifted downwards from that of the optical input. Thus, the direction of the optical frequency shift can be switched by changing the polarity of the optical bias ϕ_B or the skew ϕ.

A pair of sinusoidal signals with $\pi/2$ (90°) phase difference can be generated by using a 90° hybrid coupler. Thus, the SSB-SC modulation can be achieved by using a set up shown in Fig. 7.10. However, it would be rather difficult to change the phase difference rapidly and precisely. On the other hand, the optical phase difference (the bias ϕ_B) can be controlled precisely by an optical phase modulator. Thus, the selection of the output optical frequency from the USB and LSB can be performed by changing ϕ_B [12].

Here, we consider the SSB-SC modulation using the DPMZM as shown in Fig. 7.11, which is fabricated on an x-cut LN substrate. A pair of sinusoidal signals with 90° phase difference, described by $\cos\omega_m t$ and $\sin\omega_m t$, are applied to the two MZMs (MZ$_A$ and MZ$_B$) through the electrodes (RF$_A$ and RF$_B$). The two MZMs are set to the null-bias. The outputs of MZ$_A$ and MZ$_B$ at points P and Q in Fig. 7.11, have both the first order sidebands (USB and LSB). Figures 7.12 and 7.13

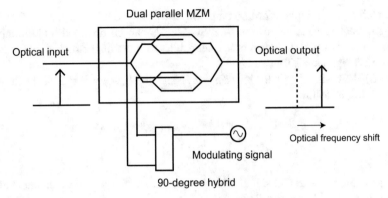

Fig. 7.10 SSB-SC modulation using a DPMZM and a 90° hybrid coupler

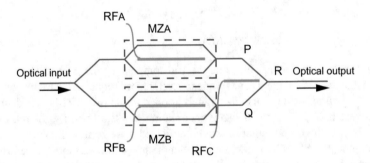

Fig. 7.11 Dual-parallel MZM

show optical spectra inside the DPMZM. The optical phase difference between the optical signals from MZ_A and MZ_B can be controlled by the voltage applied to the electrode RF_C.

When the optical phase delay induced on the output from MZ_B (Q) with respect to the phase of the output from MZ_A (P) is $\pi/2$ (+90°), the LSB is suppressed at the point R due to the destructive interference, as shown in Fig. 7.12. Thus, we can obtain the SSB-SC signal with the first order USB at the output (R). On the other hand, when the induced phase of the signal from Q is $-\pi/2$ (−90°), the modulator generates the SSB-SC signal with the first order LSB at the output (R), where the USB is suppressed by the destructive interference.

The output frequency can be switched by changing the voltage applied to the electrode RF_C. The sideband switching speed depends on the bandwidth of the electrode. The DPMZMs with traveling wave electrodes can be used for high-speed optical FSK, where the bitrate can be up to 10 Gb/s [6, 12].

Here, we derive the mathematical expression for the output of the DPMZM. The optical phase in the MZ_A induced by the modulating signal applied to the electrode

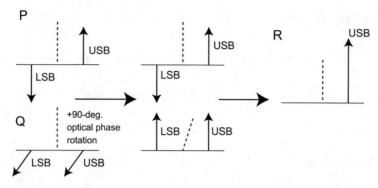

Fig. 7.12 Principle of SSB-SC (USB)

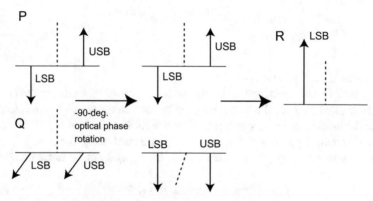

Fig. 7.13 Principle of SSB-SC (LSB)

RF$_A$ is $A_m \cos \omega_m t$, where the frequency is $f_m = \omega_m / 2\pi$. Similarly, that the optical phase in MZ$_B$ is denoted by $A_m \sin \omega_m t$. The optical signals generated by the two MZMs (MZ$_A$, MZ$_B$) are given by

$$P = \frac{1}{2\sqrt{2}} \left[e^{i(\omega_0 t + A_m \cos \omega_m t)} - e^{i(\omega_0 t - A_m \cos \omega_m t)} \right] \tag{7.59}$$

$$Q = \frac{1}{2\sqrt{2}} \left[e^{i(\omega_0 t + A_m \sin \omega_m t)} - e^{i(\omega_0 t - A_m \sin \omega_m t)} \right], \tag{7.60}$$

where P and Q denote the optical signals at the points P and Q.

P can be rewritten as,

$$P = i\frac{e^{i\omega_0 t}}{\sqrt{2}}\sin(A_m\cos\omega_m t) = \sqrt{2}ie^{i\omega_0 t}\sum_{k=0}^{\infty}(-1)^k J_{2k+1}(A_m)\cos\{(2k+1)\omega_m t\}$$

$$= \frac{i}{\sqrt{2}}e^{i\omega_0 t}\sum_{k=0}^{\infty}(-1)^k J_{2k+1}(A_m)\left\{e^{i(2k+1)\omega_m t} + e^{-i(2k+1)\omega_m t}\right\}.$$

$$(7.61)$$

Q can be expressed by

$$Q = i\frac{e^{i\omega_0 t}}{\sqrt{2}}\sin(A_m\sin\omega_m t) = \sqrt{2}ie^{i\omega_0 t}\sum_{k=0}^{\infty}J_{2k+1}(A_m)\sin\{(2k+1)\omega_m t\}$$

$$= \frac{1}{\sqrt{2}}e^{i\omega_0 t}\sum_{k=0}^{\infty}J_{2k+1}(A_m)\left\{e^{i(2k+1)\omega_m t} - e^{-i(2k+1)\omega_m t}\right\},$$

$$(7.62)$$

where we used (6.39) and (6.41).

Here, the induced optical phase in the optical waveguide connecting the points P and R is denoted by $f_{FSK}(t)$ which is induced by the voltage applied to the electrode RF_C. If the modulator has a symmetric structure for the ideal push-pull operation, the induced optical phase in the optical waveguide connecting the points Q and R can be given by $-f_{FSK}(t)$. The optical signal at the point R can be expressed by

$$R = \frac{1}{\sqrt{2}}\left[P\times e^{if_{FSK}(t)} + Q\times e^{-if_{FSK}(t)}\right].$$

$$(7.63)$$

By using (7.61) and (7.62), we get,

$$R = e^{i[\omega_0 t+\pi/4]}\bigg[\cos[f_{FSK}(t)+\pi/4]\left\{J_1(A_m)e^{i\omega_m t} - J_3(A_m)e^{-i3\omega_m t}\right.$$

$$\left. +J_5(A_m)e^{i5\omega_m t} - J_7(A_m)e^{-i7\omega_m t+\cdots}\right\}$$

$$+i\sin[f_{FSK}(t)+\pi/4]\left\{J_1(A_m)e^{-i\omega_m t} - J_3(A_m)e^{i3\omega_m t}\right.$$

$$\left. +J_5(A_m)e^{-i5\omega_m t} - J_7(A_m)e^{i7\omega_m t+\cdots}\right\}\bigg].$$

$$(7.64)$$

By neglecting the fifth or higher order sidebands, the output can be expressed approximately by

$$R = e^{i[\omega_0 t+\pi/4]}\bigg[\cos[f_{FSK}(t)+\pi/4]$$

$$\times\left\{J_1(A_m)e^{i\omega_m t} - J_3(A_m)e^{-i3\omega_m t}\right\}$$

$$+i\sin[f_{FSK}(t)+\pi/4]$$

$$\times\left\{J_1(A_m)e^{-i\omega_m t} - J_3(A_m)e^{i3\omega_m t}\right\}\bigg].$$

$$(7.65)$$

When the amplitudes of the third order sidebands can be neglected as well, the output can be expressed by the following simple form:

$$R = e^{i[\omega_0 t + \pi/4]} \Big[\cos[f_{\text{FSK}}(t) + \pi/4] J_1(A_m) e^{i\omega_m t}$$

$$+ i \sin[f_{\text{FSK}}(t) + \pi/4] J_1(A_m) e^{-i\omega_m t} \Big]. \tag{7.66}$$

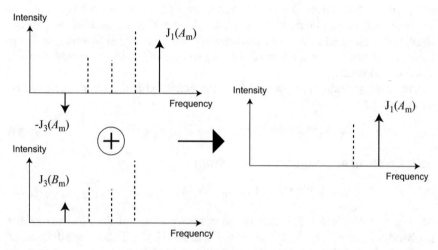

Fig. 7.14 Suppression of the third order sideband by modulation with the third order harmonic [12]

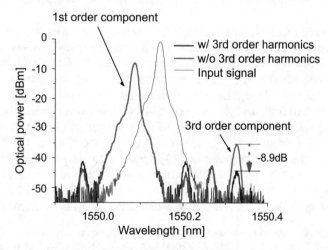

Fig. 7.15 Example of optical frequency shift by the SSB-SC [12]

When $f_{FSK}(t) = -\pi/4$, the output consists of the USB. On the other hand, when $f_{FSK}(t) = +\pi/4$, the LSB is generated at the output. The optical phase difference induced by the voltage applied to the electrode RF$_C$ is equal to $2f_{FSK}(t)$.[2] When the phase difference is equal to $+\pi/2$ or $-\pi/2$, the DPMZM generates a frequency shifted optical signal. Thus, we can generate an optical binary FSK signal by mapping $f_{FSK}(t) = -\pi/4$ as "0" and $f_{FSK}(t) = +\pi/4$ as "1".

As described in the Sect. 5.1.3, the optical vector modulation can offer the FSK modulation, where the real and imaginary components are modulated by $\cos \omega_m t$ and $\sin \omega_m t$, respectively, where the modulated signal is expressed as a temporal waveform on the phasor diagram. Here, on the other hand, the FSK or SSB-SC signals are described in the frequency domain with the Bessel function, which is useful to describe the operation of the frequency shift and the intensities of the spectral components.

The third or lower order sidebands in the output with $f_{FSK}(t) = -\pi/4$ can be written by

$$R = e^{i[\omega_0 t + \pi/4]} \left\{ J_1(A_m)e^{i\omega_m t} - J_3(A_m)e^{-i3\omega_m t} \right\}, \qquad (7.67)$$

while the output with $f_{FSK}(t) = +\pi/4$ is given by

$$R = ie^{i[\omega_0 t + \pi/4]} \left\{ J_1(A_m)e^{-i\omega_m t} - J_3(A_m)e^{i3\omega_m t} \right\}, \qquad (7.68)$$

where there are some differences in the optical phases and the polarities of the sideband components, by comparing with (7.57) and (7.58), These phase differences are due to the definition of the origin of the time and phase.

The third order term, $J_3(A_m)$, which is the lowest order distortion component due to the nonlinearity of the phase modulation, can be suppressed by feeding the third order harmonic whose frequency is $3f_m$ to the electrodes of the two MZMs RF$_A$ and RF$_B$, where the fundamental component whose frequency is f_m is simultaneously fed.

As shown in Fig. 7.14, the third order LSB is the lowest order and the dominant undesired component, when the fundamental components have $+90°$ phase difference between the two signals applied to the two electrodes. By feeding the third order harmonics of $3f_m$ with $-90°$ phase difference, the output would have an optical spectral component whose frequency is shifted by $-3f_m$. This spectral component directly generated from the $3f_m$ component in the modulating signals, and the third order LSB generated from the fundamental (f_m) component can cancel each other out, by tuning the amplitude (B_m shown in Fig. 7.14) and phase of the third order harmonics in the modulating signals [13]. Figure 7.15 shows an example of the optical spectrum of the frequency shifted optical signal. The center frequency of the output is higher than that of the input, where the frequency difference is 7.5 GHz. The output has a small distortion component, which corresponds to the third order LSB.

---

[2] The DPMZM generates the USB, when the bias $\phi_B = +\pi/2$ and the skew $\phi = +\pi/2$. The actual optical phase difference $2f_{FSK}(t)$ should be equal to $\pi/2 - \pi$, because the skew is defined with respect to the ideal push-pull operation condition. That means that $f_{FSK}(t) = -\pi/4$ is the condition for the USB signal generation.

The frequency difference between the input signal and this distortion component is 22.5 GHz, which is equal to triple the modulating-signal frequency.

Most of the 90° hybrid couplers have two inputs. The signal fed to one of the input ports is divided into two with +90° phase difference. On the other hand, the signal fed to the other input port is divided with −90° phase difference. By using this function, we can apply the fundamental and the third order harmonic components simultaneously. +90° and −90° phase differences are maintained for the fundamental and the third order harmonic components, respectively.

Problems

7.1 Approximate (7.33) by the first order term of A, for $A \ll 1$.

7.2 Derive (7.43) from (7.42).

7.3 Approximate (7.43) by the first order term of A, for $A \ll 1$.

7.4 Derive (7.64).

References

1. T. Kawanishi, *Wired and Wireless Seamless Access Systems for Public Infrastructure* (Artech House, Norwood, 2020)
2. F. Koyama, K. Iga, Frequency chirping in external modulators. J. Lightw. Technol. **6**(1), 87–93 (1988)
3. C.H. Cox III, *Analog Optical Links* (Cambridge University Press, Cambridge, 2004)
4. T. Pfau, S. Hoffmann, O. Adamczyk, R. Peveling, V. Herath, M. Porrmann, R. Noé, Coherent optical communication: towards realtime systems at 40 Gbit/s and beyond. Opt. Express **16**(2), 866–872 (2008)
5. C.R.S. Fludger, T. Duthel, D. van den Borne, C. Schulien, E. Schmidt, T. Wuth, J. Geyer, E. De Man, G. Khoe, H. de Waardt, Coherent equalization and POLMUX-RZ-DQPSK for robust 100-GE transmission. J. Lightw. Technol. **26**(1), 64–72 (2008)
6. T. Kawanishi, T. Sakamoto, M. Izutsu, High-speed control of lightwave amplitude, phase, and frequency by use of electrooptic effect. IEEE J. Sel. Topics Quant. Electron. **13**(1), 79–91 (2007)
7. T. Kawanishi, T. Sakamoto, M. Tsuchiya, M. Izutsu, S. Mori, K. Higuma, 70dB extinction-ratio Ti : LiNbO$_3$ optical intensity modulator for two-tone lightwave generation, in *Optical Fiber Communication Conference* (Optical Society of America, Washington, 2006), p. OWC4

8. K. Inagaki, T. Kawanishi, M. Izutsu, Optoelectronic frequency response measurement of photodiodes by using high-extinction ratio optical modulator. IEICE Electron. Express **9**(4), 220–226 (2012)

9. T. Kawanishi, A. Kanno, H.S.C. Freire, Wired and wireless links to bridge networks: seamlessly connecting radio and optical technologies for 5G networks. IEEE Microw. Mag. **19**(3), 102–111 (2018)

10. M. Izutsu, S. Shikamura, T. Sueta, Integrated optical SSB modulator/frequency shifter. IEEE J. Quant. Electron. **17**(11), 2225–2227 (1981)

11. T. Sakamoto, T. Kawanishi, M. Izutsu, Optical minimum-shift-keying with external modulation scheme. Opt. Express **13**, 7741–7747 (2005)

12. T. Kawanishi, K. Higuma, T. Fujita, J. Ichikawa, T. Sakamoto, S. Shinada, M. Izutsu, LiNbO$_3$ high-speed optical FSK modulator. Electron. Lett. **40**(11), 691–692 (2004)

13. T. Kawanishi, M. Izutsu, Linear single-sideband modulation for high-SNR wavelength conversion. IEEE Photon. Technol. Lett. **16**(6), 1534–1536 (2004)

Chapter 8
Estimation of Optical Modulators

This chapter describes various methods for estimation of optical modulator parameters, such as V_π, α_0, and η. As shown in Chaps. 6 and 7, the outputs of optical modulators based on the EO effect can be described by Bessel function. The mathematical model describes output spectra of actual LN optical modulators, because the refractive index change is precisely proportional to the applied voltage. The change of the absorption coefficient of the optical waveguides in the LN modulators is negligibly small. Thus, these parameters, which describe performance of the modulators, can be precisely estimated from ratios of sideband components generated by optical modulators whose structures are unknown. Firstly, this chapter shows estimation methods for a modulator consisting of a single MZI. Measurement methods for a modulator with many small MZIs are also described.

A modulator response can be also measured by electronic measuring instruments, through a photodetector which converts an optical signal into an electric signal, where frequency can be swept precisely and rapidly. Various measurement methods for frequency responses are also discussed in this chapter.

8.1 Estimation of Half-Wave Voltage and Chirp Parameter

As described in Sect. 4.3.1, the half-wave voltage of the MZM is defined by the voltage required to change the states of the MZM from the on-state to the off-state. The change should be quasi-static where frequency dependence of the modulator can be negligible.

As described in (6.114), we consider a signal with a sinusoidal waveform component proportional to $\sin \omega_m t$. The halfwave voltage is given by

$$V_{\pi \text{MZM(RF)}} = \frac{\pi V_{\text{pp}}}{2|A_1 - A_2|} = \frac{\pi V_{0\text{p}}}{|A_1 - A_2|},$$ (8.1)

© Springer Nature Switzerland AG 2022
T. Kawanishi, *Electro-optic Modulation for Photonic Networks*, Textbooks in
Telecommunication Engineering, https://doi.org/10.1007/978-3-030-86720-1_8

where the amplitude of the sinusoidal signal is V_{0p}(zero-to-peak) or V_{pp} (peak-to-peak). The optical phase difference between the two phase modulators in the MZM is varied by applying the sinusoidal signal. When the amplitude is equal to the halfwave voltage, the peak of the phase difference is equal to π.

The intrinsic chirp parameter α_0 which is defined by (6.122) can be expressed by A_1 and A_2, as well.

In this chapter, the halfwave voltage defined in Sect. 4.3.1 is denoted by $V_{\pi MZM(DC)}$, to clarify the frequency dependence. In general, the half-wave voltage for an RF signal, $V_{\pi MZM(RF)}$, is larger than $V_{\pi MZM(DC)}$, due to velocity mismatch, large Joule loss and impedance mismatch in the electrode for high-frequency signals. The halfwave voltage $V_{\pi MZM(RF)}$, depending on the frequency of the sinusoidal signal, increases rapidly in high-frequency region, so that it limits the bandwidth of the modulator.

8.1.1 Estimation of Bias Condition

The important parameters of MZMs such as $V_{\pi MZM(RF)}$, $V_{\pi MZM(DC)}$, and α_0 can be obtained by solving nonlinear simultaneous equations for A_1 and A_2, where intensities of sideband components P_n are measured by an optical spectrum analyzer. η, which describes the imbalance of the optical circuit in the MZM, and the skew, ϕ, can be also estimated from the sideband components, P_n.

The optical phase difference induced by the bias voltage applied to the electrode (ΔB) can be also obtained from the simultaneous equations.

The number of equations required to obtain these six unknown parameters is more than or equal to six. It would be rather difficult to solve simultaneous nonlinear equations in general. However, various types of mathematical methods have been developed to estimate modulator parameters under some conditions, where particular features of actual modulators are taken into account to simplify the equations. We will compare these methods to discuss details on precise measurement of modulator parameters.

The induced phases A_1 and A_2 are indispensable for estimation of the half-wave voltage and chirp parameters, which are the most important figures to describe basic characteristics of modulators. Thus, the estimation of these two parameters should be performed approximately by neglecting misalignment from the ideal balanced MZM structure. The parameters which describe the imbalance in MZMs can be calculated by using the A_1 and A_2 approximately estimated with neglecting the parameters.

When the imbalance in the optical waveguides in the MZI, and the skew in the electric signals applied to the electrodes are small enough, we can assume that $\eta = 0$ and $\phi_1 = \phi_2 = 0$. By using (6.129), the power of the n-th order sideband component is given by

$$
\begin{aligned}
P_n &= \frac{1}{4} \left| J_n(A_1) + J_n(A_2) e^{-i\phi_B} \right|^2 \\
&= \frac{1}{4} \left[\{J_n(A_1)\}^2 + \{J_n(A_2)\}^2 + 2 J_n(A_1) J_n(A_2) \cos \phi_B \right],
\end{aligned} \tag{8.2}
$$

where K is a coefficient describing the total power of the optical output from the modulator. The coefficient depends on the optical loss in the modulators, and on the sensitivity of the detector which is used for optical spectrum analysis. By neglecting the skew, the spectral shapes expressed by (8.2) would be symmetric with respect to the carrier component ($n = 0$), where $P_n = P_{-n}$. In principle, we can obtain A_1, A_2, and ϕ_B from a plural of measured sideband components P_n. However, it would be rather difficult to measure absolute values of P_n, precisely. To avoid this difficulty, we can use the power ratios of the sideband components, which are defined by

$$R_{n,m} = \frac{P_n}{P_m},$$
(8.3)

where n and m denote the orders of the sideband components.

As described in Sect. 4.3.1, most of the MZMs designed for intensity or amplitude modulation use the push-pull operation to reduce the half-wave voltages, where the induced phases A_1 and A_2 have opposite signs each other. By applying $J_n(A) = (-1)^n J_n(-A)$ to (8.2), we can obtain the following equation,

$$P_n = \frac{1}{4} \left| J_n(A) + J_n(-A)e^{-i\phi_B} \right|^2$$
$$= \frac{1}{2} \left[\{J_n(A)\}^2 (1 + (-1)^n) \cos \phi_B \right].$$
(8.4)

The even order sideband components have maximum powers at $\phi_B = 0$, and minimum powers at $\phi_B = \pi$. On the other hand, the odd order components would maximum powers at $\phi_B = \pi$, and minimum powers at $\phi_B = 0$. When the bias voltage can be controlled precisely, we can estimate the optical phase ϕ_B, where the third term of (8.2) depends on ϕ_B, sinusoidally. In this case, we can obtain the two induced phases A_1 and A_2 from two simultaneous equations, for example, for $R_{0,1}$ and $R_{0,2}$.

8.1.2 Measurement by Bias Condition Sweeping

The bias condition has fluctuation in time domain, due to DC drift or temperature change by Joule loss in the electrodes. In some MZMs, we cannot control the optical phase by an external voltage. In such cases, ϕ_B should be treated as an unknown parameter. We can obtain the values of the three parameters A_1, A_2, and ϕ_B, by using three simultaneous equations. However, nonlinear simultaneous equations have multiple solutions, and some of them are unphysical solutions. It would be difficult to select a solution with physical meaning, if the number of the equation is equal to that of the unknown parameters. In such cases, we can identify solutions related to actual physical parameters, by using equations more than the number of the unknown parameters.

The impact of the change of the optical phase difference ϕ_B can be suppressed by sweeping the voltage to control the bias condition, when the sweep range should be equal to double an even multiple of the halfwave voltage. Normally, we sweep the bias voltage with double the halfwave voltage to reduce the required voltage amplitude [1]. While the bias condition ϕ_B varies in time, the halfwave voltage is very stable. Thus, an averaged spectrum for all bias conditions, by sweeping the bias condition with a triangle wave. The period of the triangle wave should be shorter than the measurement time for the spectrum analysis, to obtain the averaged spectrum, precisely. On the other hand, the period should be much longer than that of the sinusoidal signal applied to the electrodes, in order to suppress sideband generation by the triangle wave.

By this averaging process, the third term of (8.2) is diminished, so that the sideband components can be expressed by

$$P_n = \frac{\{J_n(A_1)\}^2 + \{J_n(A_2)\}^2}{4},\qquad(8.5)$$

where all sideband components do not depend on the bias condition ϕ_B [1, 2].

This principle can be also used for measurement of the halfwave voltage $V_{\pi\text{MZM(DC)}}$. The amplitude of the sweep signal is precisely equal to an even multiple of the halfwave voltage, when the spectrum does not depend on any bias condition fluctuation due to voltage or temperature changes. The halfwave voltage $V_{\pi\text{MZM(DC)}}$ is precisely equal to the half of the minimum sweep signal amplitude which gives a stable spectrum.

8.1.3 Measurement with Null and Full Bias Conditions

The bias condition is swept for averaging or is fixed at a particular point in the measurement methods shown in the previous sections. To estimate the modulator parameters precisely, we need to measure high order sideband components, where the halfwave voltage and the chirp parameter can be calculated from A_1 and A_2 by using (6.122) and (8.1). If the half-wave voltage of the modulator is not small, the required voltage of the sinusoidal signal would be very large. It would be rather difficult to generate large amplitude high-frequency signals. Power amplifiers for high-frequency bands would be very costly.

On the other hand, by using multiple bias conditions, we can estimate the induced phases A_1 and A_2 with a relatively low amplitude stimulus sinusoidal signal for sideband generation [3].

As the first step, the optical output power is measured without any high-frequency signals for sideband generation. When no sinusoidal signal is applied to the modulator, i.e., $(A_1 = A_2 = 0)$, the optical output is expressed by

$$P_0' = \frac{1 + \cos\phi_B}{2},\qquad(8.6)$$

which goes to a maximum ($P'_{0F} = 1$) with the full bias condition ($\phi_B = 0$), and to a minimum ($P'_{0N} = 0$) with the null-bias condition. The minimum equals zero, because we assume $\eta = 0$. However, η which denotes the imbalance in the MZI should have a finite value in an actual modulator. The optical output power still has a minimum with the null-bias condition, while $P'_{0N} = 0$ would have a finite value.

As the second step, we measure the ratio of the carrier (the zeroth order) and the first sideband components, with applying the sinusoidal signal. Each component should be maximized by tuning the bias condition, where the carrier component goes to a maximum P_{0F} with the full bias condition and the first order sideband component goes to a maximum P_{1N} with the null-bias condition. Then, we can construct two equations from ratios between the maximum and minimum values of the carrier and first order sideband components, as follows:

$$\frac{P_{0F}}{P'_{0F}} = \frac{\{J_0(A_1) + J_0(A_2)\}^2}{4} \tag{8.7}$$

$$\frac{P_{1N}}{P'_{0F}} = \frac{\{J_1(A_1) - J_1(A_2)\}^2}{4}. \tag{8.8}$$

The induced phases A_1 and A_2 can be obtained by solving these equations. Finally, the halfwave voltage and the chirp parameter can be estimated from A_1 and A_2.

In this method, we use the maximum values of the sideband components where other components would be minimized. Thus, the power of the required component can be precisely measured, without applying large amplitude high-frequency signals. Figure 8.1 shows measured the half-wave voltage and chirp parameter of an LN MZM modulator, in a wide frequency range up to 40 GHz [3]. The optical spectra for various modulating frequency can be measured by a setup with an optical spectrum analyzer, a voltage source, and an RF signal source, as shown in Fig. 8.2. The sinusoidal modulating signal frequency f_m can be precisely controlled by the RF signal source, whose frequency resolution can be up to a few Hz. However, when the sinusoidal signal frequency is less than the frequency resolution of the optical spectrum analyzer, intensity of each spectral component generated by the MZM cannot be measured separately. A typical frequency resolution of an optical spectrum analyzer is a few GHz. Thus, this measurement method can be used only for high-frequency operation.

When the intensities of sideband components higher than the first order are negligibly small, the required components, P_{0F}, P'_{0F}, P_{1N}, and P'_{0F}, can be measured by an optical power meter without using any optical spectrum analyzers, to reduce the cost for the measurement equipment. This method using optical power measurement can be applied to frequency ranges lower than a few GHz. As mentioned above, the measurement with an optical spectrum analyzer can be used only for frequency ranges higher than a few GHz, due to the limitation of the frequency resolution of the optical spectrum analyzer for the sideband measurement. The bias condition would be shifted due to temperature change when the sinusoidal signal is turned on or off. Even in such cases, the required components, P_{0F}, P'_{0F}, P_{1N}, and P'_{0F} can be measured precisely, as maximums by sweeping the bias condition.

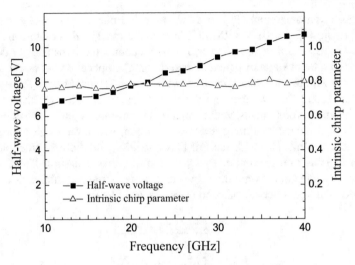

Fig. 8.1 Frequency response of halfwave voltage and chirp parameter [3]

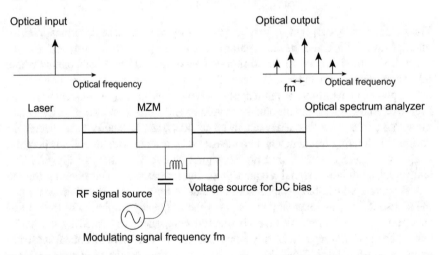

Fig. 8.2 Setup for frequency response measurement using an optical spectrum analyzer

Figure 8.3 shows another configuration of modulator parameter measurement. Intensities of sideband components are measured by a heterodyne method where the optical frequency domain profile is converted into an electric signal through signal mixing by a photodetector. Intensity of each sideband component can be precisely measured by using an RF signal spectrum analyzer, even if the modulation frequency is less than a few GHz. The frequency response of the photodetector should be measured in advance for calibration, as shown in Sect. 8.3. The frequency resolution of this measurement is dominated by the frequency stability of the lasers, where one is for the optical input and another is for a photonic local oscillator signal. The RF spectrum analyzer can measure spectral components whose frequency is less than

100 GHz, on the other hand, the optical spectrum analyzer can measure spectral components whose separation is larger than 100 GHz, Thus, the RF spectrum analyzer based measurement can be used for low-frequency region, while the optical spectrum analyzer based measurement can cover high-frequency region over 100 GHz. Figure 8.4 shows an EO frequency response of a phase modulator measured by the method shown in Fig. 8.3, where narrow linewidths lasers are used to measure response in a frequency region less than 10 kHz. The frequency response has some small dips and peaks in a frequency region from 10 MHz to 100 MHz, where AO effect is also induced by the modulating signal. Thus, the frequency response has some fluctuation due to undesired interference between the AO and EO effects.

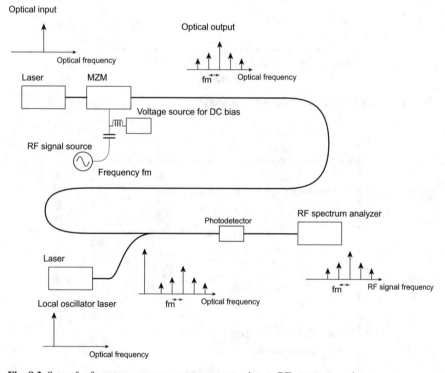

Fig. 8.3 Setup for frequency response measurement using an RF spectrum analyzer

When the modulation depth is not large enough, it would be rather difficult to measure A_1 and A_2 independently. Assuming that the induced phases are small ($|A_1|, |A_2| < 1$), the zeroth and the first order Bessel functions can be approximately expressed by

$$J_1(A_i) \sim A/2 \tag{8.9}$$

$$J_0(A_i) \sim 1, \tag{8.10}$$

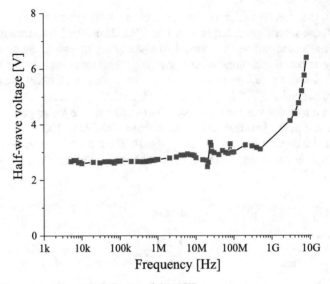

Fig. 8.4 EO response of an optical phase modulator [4]

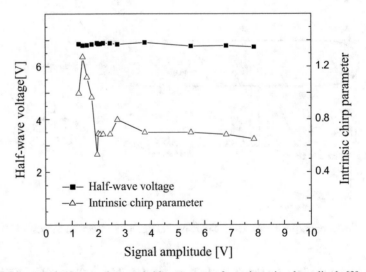

Fig. 8.5 Measured halfwave voltage and chirp parameter for various signal amplitude [3]

where $i = 0, 1$. Equations (8.7) and (8.8) can be rewritten as

$$\frac{P_{0F}}{P'_{0F}} = 1 \tag{8.11}$$

$$\frac{P_{1N}}{P'_{0F}} = \frac{(A_1 - A_2)^2}{16}. \tag{8.12}$$

$|A_1 - A_2|$ denotes the amplitude of the optical phase difference between the two arms in the MZI, where A_1 and A_2 have opposite signs each other. This equation means that $A_1 - A_2$, which is required for estimation of the half-wave voltage, can be measured precisely even if the sinusoidal signal for sideband generation is small, as shown in Fig. 8.5. The halfwave voltage does not depend on the applied signal voltage, because the EO effect in LN is precisely proportional to the applied voltage.

On the other hand, the estimated chirp parameter fluctuates when the amplitude of the applied signal is small. In principle, the chirp parameter does not depend on the applied voltage. This is due to the measurement error on $A_1 + A_2$, when the modulation depth is small.

8.2 Estimation of Parallel Mach-Zehnder Modulators

As shown in Fig. 8.6, we consider a measurement method for a parallel MZM consisting of integrated MZIs, where the half-wave voltages ($V_{\pi MZM}$), the chirp parameters (α_0), and the parameters of the interferometer imbalance (η) for all MZIs can be estimated from the sideband components in the optical output.

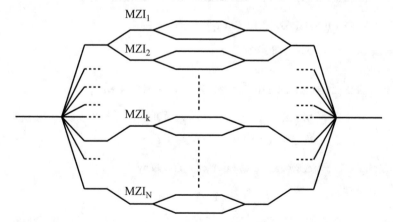

Fig. 8.6 Parallel MZM

A sinusoidal signal is applied to the MZI under measurement. Each MZI consists of two arms (optical waveguides) sandwiched between two Y-junctions, where each arm has an optical phase modulator. As we discussed, η and α_0 should be minimized to achieve ideal amplitude modulation.

When we apply a sinusoidal signal $\sin \omega_m t$ to the k-th MZI (MZI$_k$ in Fig. 8.6, we can obtain an optical output P described by

$$
\begin{aligned}
P &= \frac{K_k e^{i\omega_0 t}}{2} \sum_n e^{in\omega_m t} \left[J_n(A_{1,k}) e^{in\phi_{1,k}} e^{iB_{1,k}} \left(1 + \frac{\eta_k}{2} \right) \right. \\
&\quad \left. + J_n(A_{2,k}) e^{in\phi_{2,k}} e^{iB_{2,k}} \left(1 - \frac{\eta_k}{2} \right) \right] + G_k e^{i\omega_0 t} \\
&= \frac{K_k e^{i\omega_0 t} e^{i(n\phi_{1,k}+B_{1,k})}}{2} \sum_n e^{in\omega_m t} \left[J_n(A_k + \alpha_k^*) \left(1 + \frac{\eta_k}{2} \right) \right. \\
&\quad \left. + J_n(-A_k + \alpha_k^*) \left(1 - \frac{\eta_k}{2} \right) e^{i(n\phi_k+B_k)} \right] + G_k e^{i\omega_0 t}.
\end{aligned}
\tag{8.13}
$$

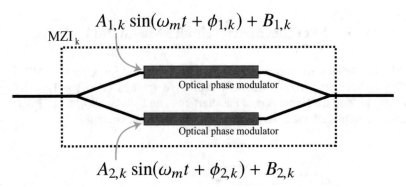

Fig. 8.7 Sinusoidal signal applied to the-k MZI

Induced optical phase at each arm in the MZI can be expressed by

$$\Phi_{1,k} = A_{1,k} \sin(\omega_m t + \phi_{1,k}) + B_{1,k} \tag{8.14}$$
$$\Phi_{2,k} = A_{2,k} \sin(\omega_m t + \phi_{2,k}) + B_{2,k}, \tag{8.15}$$

as shown in Fig. 8.7. Here, we define the induced phases by

$$A_{1,k} \equiv A_k + \alpha_k^* \tag{8.16}$$
$$A_{2,k} \equiv -A_k + \alpha_k^* \tag{8.17}$$
$$\alpha_k^* \equiv A_k \alpha_{0,k} \tag{8.18}$$
$$B_k \equiv B_{2,k} - B_{1,k} \tag{8.19}$$
$$\phi_k \equiv \phi_{2,k} - \phi_{1,k}. \tag{8.20}$$

When $A_{1,k} = -A_{2,k}$ ($\alpha_{0,k} = \alpha_k^* = 0$), we can achieve ideal push-pull operation of MZIs.

$\alpha_{0,k}$ and B_k denote the intrinsic chirp parameter and the bias condition of the k-th MZI. ϕ_k is the skew between the two sinusoidal signals applied on the two arms. If the two arms have two electrodes separately, ϕ_k can be almost zero by using an external driving circuit to compensate fabrication error in the MZI. η_k is the optical power difference between the arms. K_k denotes the total insertion loss of MZI_k.

The optical output would include unmodulated optical signals from other MZIs. The intensities and phases of the unmodulated signals which depend on the bias conditions of the MZIs, $B_{1,j}, B_{2,j}(j \neq k)$, can be expressed by

$$G_k = \sum_{j=1}^{N} \frac{K_j}{2} \left[e^{iB_{1,j}} \left(1 + \frac{\eta_j}{2}\right) + e^{iB_{2,j}} \left(1 - \frac{\eta_j}{2}\right)\right]$$

$$- \frac{K_k}{2} \left[e^{iB_{1,k}} \left(1 + \frac{\eta_k}{2}\right) + e^{iB_{2,k}} \left(1 - \frac{\eta_k}{2}\right).\right] \qquad (8.21)$$

When the impact of the skew can be neglected, i.e., $\phi_k = 0$, the n-th order sideband intensity $P_{n,k}(n \neq 0)$ can be expressed by

$$P_{n,k} = \frac{K_k^2}{4} \left| J_n(A_k + \alpha_k^*)\left(1 + \frac{\eta_k}{2}\right) + J_n(-A_k + \alpha_k^*)\left(1 - \frac{\eta_k}{2}\right) e^{iB_k} \right|^2$$

$$= \frac{K_k^2}{4} \left[J_n^2(A_k + \alpha_k^*)\left(1 + \frac{\eta_k}{2}\right)^2 + J_n^2(-A_k + \alpha_k^*)\left(1 - \frac{\eta_k}{2}\right)^2 \right.$$

$$\left. + 2\cos B_k J_n(A_k + \alpha_k^*) J_n(-A_k + \alpha_k^*)\left(1 - \frac{\eta_k^2}{4}\right) \right]. \qquad (8.22)$$

The terms of $e^{in\omega_m t}$ in (8.13) correspond to the n-th order sideband components, whose optical frequency is $(\omega_0 + n\omega_m)/2\pi$. The zeroth order components, which include the optical signals from other MZIs (G_k), are not used for estimation of the parameters of the MZI under measurement (MZI_k). As shown in Fig. 8.8, by excluding the zeroth order component, the parameters can be measured without any impact from the bias conditions of the other MZIs [5]. Thus, we can focus on the control of the bias and the sinusoidal signal applied on the MZI under measurement (MZI_k).

When we neglect the effect of the skew, $P_{n,k} = P_{-n,k}$. Thus, we can calculate the required parameters only from USB or LSB. On the other hand, we can judge if the skew is small enough or not, by monitoring the difference between $P_{n,k}$ and $P_{-n,k}$.

For simplicity, we assume that $|\alpha_{0,k}|, |\eta_k| \ll 1$. When $B_k = 0$, the odd order sideband components have minimum values, and the even order sideband components have maximum values. When $B_k = \pi$, the relation is vice versa. Thus, the intensities of the odd and even order sideband components oscillate between the maximum and minimum values, by sweeping the bias voltage continuously. Thus, we can find the bias conditions for $B_k = 0$ and $B_k = \pi$, by monitoring the optical spectrum with bias condition sweeping.

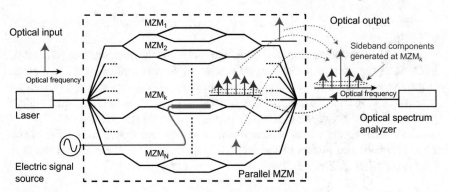

Fig. 8.8 Sideband generation from an MZI in a parallel MZM [5]

We can construct nonlinear simultaneous equations for A_k, α_k^*, η_k, by using sideband components measured by the two bias conditions ($B_k = 0$ and $B_k = \pi$).

Here, $P_{n,k}^{(-)}$ denotes the optical power of the even order sideband components with the bias conditions of $B_k = 0$, or that of the odd order sideband components with the bias conditions of $B_k = \pi$. On the other hand, $P_{n,k}^{(+)}$ is the optical power of the even order sideband components with the bias conditions of $B_k = \pi$, or that of the odd order sideband components with the bias conditions of $B_k = 0$. In other words, the maximum and minimum values of the n-th order sideband component with the bias condition sweeping are, respectively, described by $P_{n,k}^{(+)}$ and $P_{n,k}^{(-)}$.

For the bias condition of $B_k = 0$, the first order sideband components can be given by

$$
\begin{aligned}
P_{1,k}^{(-)} &= \frac{K_k^2}{4} \left[J_1(A_k + \alpha_k^*)\left(1 + \frac{\eta_k}{2}\right) - J_1(A_k - \alpha_k^*)\left(1 - \frac{\eta_k}{2}\right) \right]^2 \\
&= \frac{K_k^2}{4} \left[\eta_k \frac{J_1(A_k + \alpha_k^*) + J_1(A_k - \alpha_k^*)}{2} + J_1(A_k + \alpha_k^*) - J_1(A_k - \alpha_k^*) \right]^2 .
\end{aligned}
$$
(8.23)

By assuming that $\alpha_{0,k} \ll 1$, we can obtain

$$
\frac{J_1(A_k + \alpha_k^*) + J_1(A_k - \alpha_k^*)}{2} \simeq J_1(A_k),
$$
(8.24)

and

$$
J_1(A_k + \alpha_k^*) - J_1(A_k - \alpha_k^*) \simeq 2\alpha_k^* J_1'(A_k),
$$
(8.25)

where $J_n'(A_k)$ is the derivative of $J_n(A_k)$. By using these approximate equations, $P_{1,k}^{(-)}$ can be described by

$$P_{1,k}^{(-)} \simeq \frac{K_k^2}{4} \left[\eta_k J_1(A_k) + 2\alpha_k^* J_1'(A_k) \right]^2 .$$ (8.26)

Similarly, the second order component can be given by

$$P_{2,k}^{(+)} = \frac{K_k^2}{4} \left[J_2(A_k + \alpha_k^*) \left(1 + \frac{\eta_k}{2} \right) + J_2(A_k - \alpha_k^*) \left(1 - \frac{\eta_k}{2} \right) \right]^2$$

$$= \frac{K_k^2}{4} \left[J_2(A_k + \alpha_k^*) + J_2(A_k - \alpha_k^*) + \eta_k \frac{J_2(A_k + \alpha_k^*) - J_2(A_k - \alpha_k^*)}{2} \right]^2$$

$$\simeq \frac{K_k^2}{4} \left[2J_2(A_k) + \alpha_k^* \eta_k J_2'(A_k) \right]^2 .$$ (8.27)

When $|\alpha_k^* \eta_k| \ll 1$, $P_{2,k}^{(+)}$ can be approximated by

$$P_{2,k}^{(+)} \simeq K_k^2 \left[J_2(A_k) \right]^2 .$$ (8.28)

For n-th order sideband components, the maximum and minimum values can be expressed by

$$P_{n,k}^{(+)} \simeq \frac{K_k^2}{4} \left[2J_n(A_k) + \alpha_k^* \eta_k J_n'(A_k) \right]^2 \simeq K_k^2 \left[J_n(A_k) \right]^2$$ (8.29)

$$P_{n,k}^{(-)} \simeq \frac{K_k^2}{4} \left[\eta_k J_n(A_k) + 2\alpha_k^* J_n'(A_k) \right]^2 .$$ (8.30)

In general, the intensity of the high order sideband components are smaller than the lower order components, so that we use the sideband components where the order number is not much larger than a few. Here, we describe an example of the procedure for estimation of A_k, α_k^* and η_k. Firstly, we can obtain A_k by substituting the measured values of $P_{1,k}^{(+)}$ and $P_{2,k}^{(+)}$ into

$$\frac{P_{2,k}^{(+)}}{P_{1,k}^{(+)}} \simeq \left[\frac{J_2(A_k)}{J_1(A_k)} \right]^2 .$$ (8.31)

If we can measure higher order sideband components, for example, the third order components, we can also calculate A_k by

$$\frac{P_{3,k}^{(+)}}{P_{1,k}^{(+)}} \simeq \left[\frac{J_3(A_k)}{J_1(A_k)} \right]^2 ,$$ (8.32)

or by

$$\frac{P_{3,k}^{(+)}}{P_{2,k}^{(+)}} \simeq \left[\frac{J_3(A_k)}{J_2(A_k)} \right]^2 .$$ (8.33)

By comparing the solutions from these equations, we can estimate preciseness of the measurement for $P_{1,k}^{(+)}$, $P_{2,k}^{(+)}$

The parameters which describe the imperfection of the MZI (η_k and $\alpha_{0,k}$) can be calculated from $P_{n,k}^{(+)}$ and $P_{n,k}^{(-)}$, A_k, n, η_k and α_k^* approximately satisfy

$$
\frac{P_{n,k}^{(-)}}{P_{n,k}^{(+)}} \simeq \left\{ \frac{\eta_k J_n(A_k) + 2\alpha_k^* J_n'(A_k)}{2J_n(A_k) + \alpha_k^* \eta_k J_n'(A_k)} \right\}^2 \simeq \left\{ \frac{\eta_k J_n(A_k) + 2\alpha_k^* J_n'(A_k)}{2J_n(A_k)} \right\}^2
$$

$$
= \left\{ \frac{\eta_k}{2} + \alpha_k^* \frac{J_n'(A_k)}{J_n(A_k)} \right\}^2 = \left[\frac{\eta_k}{2} + \alpha_k^* \left\{ \frac{J_{n-1}(A_k)}{J_n(A_k)} - \frac{n}{A_k} \right\} \right]^2 . \tag{8.34}
$$

Equations of $P_{n,k}^{(-)}$ and $P_{m,k}^{(+)}$ ($n \neq m$) can be also used for estimation of η_k and α_k^*

For example, to estimate η_k and $\alpha_{0,k}$ we can construct simultaneous equations for η_k and α_k^*, as follows,

$$
\frac{P_{1,k}^{(-)}}{P_{1,k}^{(+)}} \simeq \left[\frac{\eta_k}{2} + \alpha_k^* \left\{ \frac{J_0(A_k)}{J_1(A_k)} - \frac{1}{A_k} \right\} \right]^2 \tag{8.35}
$$

$$
\frac{P_{2,k}^{(-)}}{P_{2,k}^{(+)}} \simeq \left[\frac{\eta_k}{2} + \alpha_k^* \left\{ \frac{J_1(A_k)}{J_2(A_k)} - \frac{2}{A_k} \right\} \right]^2 \tag{8.36}
$$

by using the measured values of $P_{1,k}^{(+)}$, $P_{1,k}^{(-)}$, $P_{2,k}^{(+)}$, and $P_{2,k}^{(-)}$.

By taking the square root of both sides, these equations can be converted into linear simultaneous equation, so that the solution can be easily estimated. However, they consist of the following four equations,

$$
\frac{\eta_k}{2} + \alpha_k^* \left\{ \frac{J_0(A_k)}{J_1(A_k)} - \frac{1}{A_k} \right\} = \pm \sqrt{\frac{P_{1,k}^{(-)}}{P_{1,k}^{(+)}}} \tag{8.37}
$$

$$
\frac{\eta_k}{2} + \alpha_k^* \left\{ \frac{J_1(A_k)}{J_2(A_k)} - \frac{2}{A_k} \right\} = \pm \sqrt{\frac{P_{2,k}^{(-)}}{P_{2,k}^{(+)}}}, \tag{8.38}
$$

because the right sides have four combinations of the signs.

If we flip the signs of (8.37) and (8.38), simultaneously, the signs of η_k and α_k^* are also flipped. Even in the equations for the higher order sideband components, the equations have difference only in the term of α_k^*, which is $J_n(A_k)/J_n(A_k) - n/A_k$, so that we cannot fix the signs of η_k and α_k^* from the maximum and minimum values of the sideband power ($P_{n,k}^{(-)}$ and $P_{n,k}^{(+)}$).

On the other hand, we can distinguish if η_k and α_k^* have the same sign or not. In other word, the sign of $\eta_k \alpha_k^*$ can be fixed. Thus, we should solve a pair of simultaneous equations with the following two conditions: (1) the right sides of (8.37) and (8.38) have the same sign, (2) they have the opposite signs. By using equations for higher order sideband components, we can select solutions with physical meanings.

For example, we fix the sign of the right side of (8.37) to positive, while that of (8.38) is positive or negative. Thus, we should solve two types of simultaneous equations for η_k and α_k^*, where A_k would be fixed with the value obtained from $P_{1,k}^{(+)}$ and $P_{2,k}^{(+)}$. We can estimate the accuracy of the calculation and measurement by using equations consisting of $P_{3,k}^{(+)}$ and $P_{3,k}^{(-)}$. The sign of A_k only depends on the definition of the polarity of the signal applied to the arms. Thus, we can fix the sign to positive without loss of generality. Solutions of (8.37) and (8.38) are substituted into the following equation:

$$\frac{P_{3,k}^{(-)}}{P_{3,k}^{(+)}} \simeq \left[\frac{\eta_k}{2} + \alpha_k^* \left\{ \frac{J_2(A_k)}{J_3(A_k)} - \frac{3}{A_k} \right\} \right]^2. \tag{8.39}$$

The physically relevant solutions should satisfy this equation.

In summary, the parameters of the k-th MZI can be estimated by the following steps. 1) measuring sideband intensities up to the third order ($P_{1,k}^{(+)}$, $P_{1,k}^{(-)}$, $P_{2,k}^{(+)}$, $P_{2,k}^{(-)}$, $P_{3,k}^{(+)}$ and $P_{3,k}^{(-)}$), 2) solving (8.31) for A_k, and 3) solving (8.35) and (8.36) for η_k and α_k^* with substituting A_k.

To improve the accuracy of η_k, $\alpha_{0,k}$ and A_k, we can construct nonlinear simultaneous equations consisting of

$$P_{n,k}^{(-)} = \frac{K_k^2}{4} \left[\eta_k \frac{J_n(A_k + \alpha_k^*) + J_n(A_k - \alpha_k^*)}{2} \right.$$
$$\left. + J_n(A_k + \alpha_k^*) - J_n(A_k - \alpha_k^*) \right]^2 \tag{8.40}$$

$$P_{n,k}^{(+)} = \frac{K_k^2}{4} \left[J_n(A_k + \alpha_k^*) + J_n(A_k - \alpha_k^*) \right.$$
$$\left. + \eta_k \frac{J_n(A_k + \alpha_k^*) - J_n(A_k - \alpha_k^*)}{2} \right]^2, \tag{8.41}$$

where the approximation described by (8.24) and (8.25) are not applied. The approximate solutions of η_k, $\alpha_{0,k}$, and A_k, obtained by the approximate equations (8.31), (8.35) and (8.36), can be used as initial values for iterative solution process of the nonlinear simultaneous equations.

8.3 Frequency Response Measurement

One of the most important parameters which describe performance of optoelectronic devices including optical modulators is operation bandwidth measured by frequency domain measurement. As described in Sect. 8.1.3, frequency responses can be measured by optical spectrum analysis. Various parameters can be obtained from optical spectra, however, the measurement speed is limited by that of the optical spectrum analyzer. In this method, the optical frequency for optical spectrum analysis is swept in the optical spectrum analyzer for a modulating signal frequency f_m, which should be also swept for frequency response measurement. In other words, two-dimensional frequency sweep is required to obtain a frequency response curve.

For quality control in fabrication or diagnosis in development of modulators, frequency response measurement based on frequency sweep of electric signals is commonly used as a rapid measurement method with a simple setup as shown in Fig. 8.9. The setup consists of an RF signal generator, a photodetector, and an RF signal receiver, where the frequency response of the MZM can be obtained from the ratio between the RF power at the MZM input port and the RF power at the photodetector output. A vector network analyzer can offer frequency response measurement with precise calibration to exclude impact of the frequency responses of the RF signal generator, cables, and detectors [6, 7]. However, the frequency response of the photodetector cannot be calibrated by measurement in electric domain. The conversion efficiency and its frequency dependency should be measured in advance by using a reference optical signal.

Here, we consider a method for photodetector calibration. The output of the photodetector current can be described by

$$I(f_m) = D(f_m)S(f_m), \tag{8.42}$$

where $D(f_m)$ denotes the photodetector conversion efficiency for a sinusoidal signal of a frequency denoted by f_m. The output electric power can be expressed by

$$P_{\text{RF}}(f_m) = \left\{ \frac{I(f_m)}{\sqrt{2}} \right\}^2 Z_L, \tag{8.43}$$

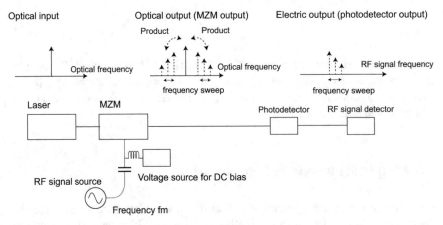

Fig. 8.9 Setup for frequency response measurement using electric domain frequency sweep

where Z_L is the load impedance of the photodetector electric output [8]. $S(f_m)$ denotes the intensity of the spectral component whose frequency is f_m in the optical power described by $|R|^2$. The intensity of the spectral component for the quadrature bias condition, can be derived from (7.17), as follows:

$$S(f_m) = 2A(f_m), \tag{8.44}$$

where $A(f_m)$ denotes the amplitude of the optical phase change induced by the modulating signal whose frequency is $f_m(= \omega_m/2\pi)$. Thus, the frequency response described by

$$I(f_m) = 2D(f_m)A(f_m) \tag{8.45}$$

includes both the frequency responses of the MZM and photodetector.

Figure 8.10 shows a configuration of photodetector calibration by using an optical two-tone signal generated by an MZM with null-bias point [9]. As described in Sect. 7.1.2, a pure optical two-tone signal can be generated by using a balanced MZM, where the frequency separation between the two spectral components is $2f_m$. The intensity of the $2f_m$ component in $|R|^2$ is

$$S(2f_m) = \frac{1}{2}\{A(2f_m)\}^2, \tag{8.46}$$

which can be derived from (7.26). The intensity of the optical two-tone signal depends on the modulating signal frequency. However, the spectral shape does not depend on the frequency, because it has only two balanced sideband components. As shown in Fig. 8.10, the intensity change due to the frequency sweep can be compensated by using an optical amplifier with automatic power level control. When the amplifier output is stabilized at P_{opt}, the amplitude of the first USB and LSB components is equal to $\sqrt{P_{opt}/2}$. The optical intensity of the $2f_m$ component can be approximately expressed by

$$S(2f_m) = P_{opt}, \tag{8.47}$$

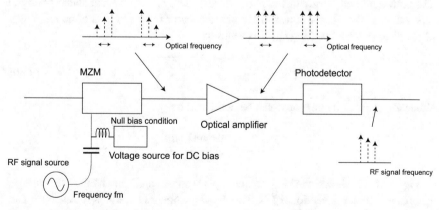

Fig. 8.10 Setup for photodetector calibration using an optical two-tone signal

which does not depend on the MZM frequency response. Thus, the photodetector output, in the configuration shown in Fig. 8.10, directly associated with the frequency response of the photodetector, as follows:

$$I(2f_m) = P_{opt}D(2f_m).$$ (8.48)

Thus, the photodetector response $D(f_m)$ can be obtained from the photodetector output without knowing the MZM frequency response $A(f_m)$. By substituting $D(f_m)$ to (8.45), we can obtain the frequency response of the MZM, $A(f_m)$. This is a basic principle of the photodetector calibration using an optical two-tone signal. When the MZM for the optical two-tone signal generation is not well-balanced, in other words, the ER is not large enough, the automatic power level control induces error due to residual carrier component. High ER modulation is important to ensure precise photodetector calibration.

When the bias of the MZM is set to a quadrature bias condition, the spectral shape largely depends on the modulating signal frequency as shown in Fig. 8.9. The intensity of the carrier component does not depend on the amplitude of the optical shift when we neglect the second or higher order terms, where the carrier would be almost constant with respect to the modulating signal frequency. Thus, an MZM output with a quadrature bias cannot be used as a reference signal for photodetector calibration without knowing $A(f_m)$.

Two-tone signals can be also generated by a pair of stable laser sources, where the total optical power can be larger than that of a two-tone signal generated by an MZM. However, it would be rather difficult to suppress fluctuation between two lasers. Polarization fluctuation, frequency excursion and amplitude fluctuation should be suppressed for accurate calibration. On the other hand, an MZM can generate a stable optical two-tone signal with a simple control circuit [10]. The output has some common fluctuation due to the optical input. However, the relative polarization difference, frequency shift, and amplitude difference are highly stable in the sideband components of the MZM output.

Figure 8.11 shows another configuration for photodetector calibration [8, 9, 11]. The conversion efficiency of the photodetector $D(2f_m)$ can be calculated from the ratio between the optical power of the photodetector input (P_{opt}) and the RF power of the photodetector output (P_{RF}), as follows:

$$D(2f_m) = \sqrt{\frac{2}{Z_L}} \frac{\sqrt{P_{RF}(2f_m)}}{P_{opt}(2f_m)}.$$ (8.49)

When Z_L is 50 Ω, the response can be described by

$$D(2f_m) = \frac{\sqrt{P_{RF}(2f_m)}}{5P_{opt}(2f_m)}.$$ (8.50)

Figure 8.12 shows the EO frequency response and the half-wave voltage, $V_{\pi MZM(RF)}$, of a conventional LN MZM, where the EO frequency response can be calculated from the half-wave voltage. When the modulating signal frequency

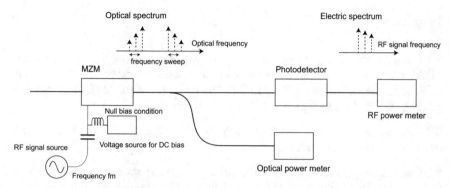

Fig. 8.11 Setup for photodetector calibration by optical power and RF power measurement

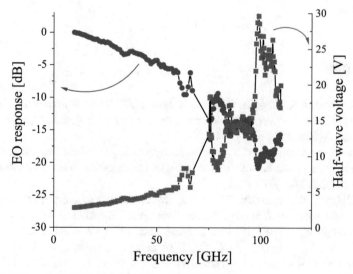

Fig. 8.12 EO response and half-wave voltage of an MZM [12]

is less than 67 GHz, the half-wave voltage was measured by using a setup shown in Fig. 8.9. On the other hand, that for the frequency range higher than 67 GHz was measured by an optical spectrum analyzer as shown in Fig. 8.2. The measurement method based on optical spectrum analysis can be used for frequency range of 50 GHz or higher, while electric domain frequency sweep can offer rapid and stable measurement in lower frequency regions.

Problems

8.1 Consider the estimation of the half-wave voltage, $V_{\pi MZM(RF)}$, of a zero-chirp MZM, where $\alpha_0 = 0$. Assume that the amplitude of the optical phase shift induced by a modulating signal is equal to 1.60 rad on each arm of the MZM, and that the power of the modulating signal is 26 dBm. Calculate $V_{\pi MZM(RF)}$.

8.2 Calculate the half-wave voltage of each phase modulator embedded in the MZM described in the previous problem.

8.3 Calculate $V_{\pi MZM(RF)}$ of an MZM whose intrinsic chirp parameter $\alpha_0 = 1$. Assume that the amplitude of the optical phase shift induced by a modulating signal is equal to 1.60 rad on one of arms of the MZM, and that the power of the modulating signal is 23 dBm.

References

1. T. Kawanishi, K. Kogo, S. Oikawa, M. Izutsu, Direct measurement of chirp parameters of high-speed Mach-Zehnder-type optical modulators. Opt. Commun. **195**(5–6), 399–404 (2001)
2. Measurement method of a half-wavelength voltage and a chirp parameter for Mach-Zehnder optical modulator in high-frequency radio on fibre (RoF) systems, IEC 62802:2017
3. S. Oikawa, T. Kawanishi, M. Izutsu, Measurement of chirp parameters and halfwave voltages of Mach-Zehnder-type optical modulators by using a small signal operation. IEEE Photon. Technol. Lett. **15**(5), 682–684 (2003)
4. J. Mochizuki, K. Inagaki, G.W. Lu, T. Sakamoto, Y. Yoshida, A. Kanno, N. Yamamoto, T. Kawanishi, Variable-linewidth light source based on Brownian motion random walk in optical phase, in *IEICE Electronics Express*, advpub, 2020
5. T. Kawanishi, Parallel Mach-Zehnder modulators for quadrature amplitude modulation. IEICE Electron. Express **8**(20), 1678–1688 (2011)
6. T. Kawanishi, *Wired and Wireless Seamless Access Systems for Public Infrastructure* (Artech House, London, 2020)
7. S. Iezekiel, Measurement of microwave behavior in optical links. IEEE Microwave Mag. **9**(3), 100–120 (2008)
8. K. Inagaki, T. Kawanishi, M. Izutsu, Optoelectronic frequency response measurement of photodiodes by using high-extinction ratio optical modulator. IEICE Electron. Express **9**(4), 220–226 (2012)
9. Transmitting equipment for radiocommunication - Frequency response of optical-to-electric conversion device in high-frequency radio over fibre systems - Measurement method, IEC 62803:2016
10. H. Kiuchi, T. Kawanishi, M. Yamada, T. Sakamoto, M. Tsuchiya, J. Amagai, M. Izutsu, High extinction ratio Mach-Zehnder modulator applied to a highly

stable optical signal generator. IEEE Trans. Microw. Theory Tech. **55**(9), 1964–1972 (2007)

11. T. Kawanishi, Precise optical modulation and its application to optoelectronic device measurement. Photonics **8**(5), 160 (2021)

12. P.T. Dat, Y. Yamaguchi, K. Inagaki, M. Motoya, S. Oikawa, J. Ichikawa, A. Kanno, N. Yamamoto, T. Kawanishi, Transparent fiber-radio-fiber bridge at 101 GHz using optical modulator and direct photonic down-conversion, in *Optical Fiber Communication Conference* (Optical Society of America, Washington, 2021), p. F3C.4

Appendix A
Composite Trigonometric Functions

By using the addition theorem of trigonometric functions, $\cos(A+B) = \cos A \cos B - \sin A \sin B$, with the Jacobi-Anger expansion described in (6.38), we can obtain an expansion for $\cos(z \sin \theta + \beta)$, as follows:

$$\cos(z \sin \theta + \beta) = \cos \beta \cos(z \sin \theta) - \sin \beta \sin(z \sin \theta)$$

$$= \cos \beta \left[J_0(z) + 2 \sum_{n=1}^{\infty} J_{2n}(z) \cos(2n\theta) \right]$$

$$- \sin \beta \left[2 \sum_{n=1}^{\infty} J_{2n+1}(z) \sin\{(2n+1)\theta\} \right]$$

$$= J_0(z) \cos \beta + \sum_{n=1}^{\infty} J_{2n}(z) \{\cos(\beta + 2n\theta) + \cos(\beta - 2n\theta)\}$$

$$- \sum_{n=1}^{\infty} J_{2n+1}(z) \{-\cos[\beta + (2n+1)\theta] + \cos[\beta - (2n+1)\theta]\}$$

$$= \sum_{n=-\infty}^{\infty} J_n(z) \cos(\beta + n\theta), \tag{A.1}$$

where we used $J_{-n}(z) = (-1)^n J_n(z)$.

Similarly, we can derive an expansion for $\cos(z \cos \theta)$, using (6.40), as follows:

$$\cos(z \cos \theta + \beta) = \cos \beta \cos(z \cos \theta) - \sin \beta \sin(z \cos \theta)$$

$$= \cos \beta \left[J_0(z) + 2 \sum_{n=1}^{\infty} (-1)^n J_{2n}(z) \cos(2n\theta) \right]$$

$$- \sin \beta \left[2 \sum_{n=1}^{\infty} (-1)^n J_{2n+1}(z) \cos\{(2n+1)\theta\} \right]$$

© Springer Nature Switzerland AG 2022
T. Kawanishi, *Electro-optic Modulation for Photonic Networks*, Textbooks in Telecommunication Engineering, https://doi.org/10.1007/978-3-030-86720-1

$$= J_0(z) \cos \beta$$

$$+ \sum_{n=1}^{\infty} (-1)^n J_{2n}(z) \left\{ \cos(\beta + 2n\theta) + \cos(\beta - 2n\theta) \right\}$$

$$- \sum_{n=1}^{\infty} (-1)^n J_{2n+1}(z) \left\{ \sin[\beta + (2n + 1)\theta] \right.$$

$$\left. + \sin[\beta - (2n + 1)\theta] \right\} . \tag{A.2}$$

By using $\sin(A + B) = \sin A \cos B + \cos A \sin B$ with (6.39) and (6.41), expansions for $\sin(z \sin \theta + \beta)$ and $\sin(z \sin \theta + \beta)$ can be derived as follows:

$$\sin(z \sin \theta + \beta) = \cos \beta \sin(z \sin \theta) + \sin \beta \cos(z \sin \theta)$$

$$= \cos \beta \left[2 \sum_{n=1}^{\infty} J_{2n+1}(z) \sin\{(2n + 1)\theta\} \right]$$

$$+ \sin \beta \left[J_0(z) + 2 \sum_{n=1}^{\infty} J_{2n}(z) \cos(2n\theta) \right]$$

$$= J_0(z) \sin \beta + \sum_{n=1}^{\infty} J_{2n}(z) \left\{ \sin(\beta + 2n\theta) + \sin(\beta - 2n\theta) \right\}$$

$$+ \sum_{n=1}^{\infty} J_{2n+1}(z) \left\{ \sin[\beta + (2n + 1)\theta] - \sin[\beta - (2n + 1)\theta] \right\}$$

$$= \sum_{n=-\infty}^{\infty} J_n(z) \sin(\beta + n\theta), \tag{A.3}$$

and

$$\sin(z \cos \theta + \beta) = \cos \beta \sin(z \cos \theta) + \sin \beta \cos(z \cos \theta)$$

$$= \cos \beta \left[2 \sum_{n=1}^{\infty} (-1)^n J_{2n+1}(z) \cos\{(2n + 1)\theta\} \right]$$

$$- \sin \beta \left[J_0(z) + 2 \sum_{n=1}^{\infty} (-1)^n J_{2n}(z) \cos(2n\theta) \right]$$

$$= J_0(z) \sin \beta$$

$$+ \sum_{n=1}^{\infty} (-1)^n J_{2n}(z) \left\{ \sin[\beta + 2n\theta] \right.$$

$$\left. + \sin[\beta - 2n\theta] \right\}$$

$$- \sum_{n=1}^{\infty} (-1)^n J_{2n+1}(z) \left\{ \cos[\beta + (2n + 1)\theta] \right.$$

$$\left. + \cos[\beta - (2n + 1)\theta] \right\} . \tag{A.4}$$

By using (A.1) and (A.3), we can get the expansion for the phasor composite function, as follows,

$$
\begin{aligned}
\exp\left[i(z\sin\theta+\beta)\right] &= \cos(z\sin\theta+\beta)+i\sin(z\sin\theta+\beta) \\
&= \sum_{n=-\infty}^{\infty} J_n(z)\left[\cos(\beta+n\theta)+i\sin(\beta+n\theta)\right] \\
&= \sum_{n=-\infty}^{\infty} J_n(z)\exp\left[i(\beta+n\theta)\right].
\end{aligned} \tag{A.5}
$$

Similarly, from (A.2) and (A.4), we can derive

$$
\begin{aligned}
\exp\left[i(z\cos\theta+\beta)\right] &= \cos(z\cos\theta+\beta)+i\sin(z\cos\theta+\beta) \\
&= J_0(z)[\cos\beta+i\sin\beta] \\
&\quad +\sum_{n=1}^{\infty}\Big[(-1)^n J_{2n}(z) \\
&\quad \times\{\cos(\beta+2n\theta)+i\sin(\beta+2n\theta) \\
&\quad +\cos(\beta-2n\theta)+i\sin(\beta-2n\theta)\}\Big] \\
&\quad +\sum_{n=1}^{\infty}\Big[(-1)^n J_{2n+1}(z) \\
&\quad \times\{[i\cos\{\beta+(2n+1)\theta\}-\sin\{\beta+(2n+1)\theta\} \\
&\quad +[i\cos\{\beta-(2n+1)\theta\}-\sin\{\beta-(2n+1)\theta\}\}\Big] \\
&= J_0(z)e^{i\beta}+\sum_{n=1}^{\infty} i^{2n} J_{2n}(z)[e^{i[\beta+2n\theta]}+e^{i[\beta-2n\theta]}] \\
&\quad +\sum_{n=1}^{\infty} i^{2n+1} J_{2n+1}(z)[e^{i[\beta+(2n+1)\theta]}+e^{i[\beta-(2n+1)\theta]}] \\
&= \sum_{n=-\infty}^{\infty} J_n(z)i^n \exp[i(\beta+n\theta)],
\end{aligned} \tag{A.6}
$$

where we used $i^{-(2n+1)}=-i^{(2n+1)}$ and $J_{-(2n+1)}(z)=-J_{2n+1}(z)$. These equations can be also derived by replacing θ with $\theta+\pi/2$ in (A.5), as follows,

$$
\begin{aligned}
\exp[i(z\cos\theta+\beta)] &= \exp[i\{z\sin(\theta+\pi/2)+\beta\}] \\
&= \sum_{n=\infty}^{\infty} J_n(z)\exp[i\{n(\theta+\pi/2)+\beta\}] \\
&= \sum_{n=-\infty}^{\infty} J_n(z)i^n \exp[i(\beta+n\theta)].
\end{aligned} \tag{A.7}
$$

Solutions

Problems of Chap. 2

2.1 The baud rate can be calculated by using (2.1) as follows:

$$f_p = \frac{1}{100 \times 10^{-12}} = 10 \times 10^9 = 10 \text{ Gbaud.}$$

For a binary modulation format, the number of symbols $N_p = 2$. By substituting them into (2.2), we get

$$C = 10 \times 10^9 \log_2 2 = 10 \text{ Gb/s.}$$

2.2 The number of symbols $N_p = 64$. The baud rate $f_p = 50 \times 10^9$. By substituting them into (2.2), we get

$$C = 50 \times 10^9 \log_2 64 = 300 \text{ Gb/s.}$$

2.3 By substituting $\lambda = 1550$ and $\Delta f = 50$ to (2.27), we get

$$\Delta\lambda = 3.3 \times 10^{-9} \times 1550^2 \times 50 = 0.396 \text{ nm.}$$

2.4 By substituting $\lambda = 1000$ and $\Delta f = 100$ to (2.27), we get

$$\Delta\lambda = 3.3 \times 10^{-9} \times 1000^2 \times 100 = 0.330 \text{ nm.}$$

Bandwidth of a 100-GHz channel at 1000 nm is narrower than that of a 50-GHz channel at 1550 nm, if we measure them in wavelength.

© Springer Nature Switzerland AG 2022
T. Kawanishi, *Electro-optic Modulation for Photonic Networks*, Textbooks in Telecommunication Engineering, https://doi.org/10.1007/978-3-030-86720-1

2.5 By using $f = c/\lambda$, the lower and upper edges of the T-band in wavelength can be converted into frequencies, where c is the speed of light, 299792458 m/s.

Edges of the T-band in frequency are

$$299792458/1000 \times 10^{-9} = 299.792 \times 10^{12} = 299.792 \text{ THz},$$

and

$$299792458/1260 \times 10^{-9} = 237.931 \times 10^{12} = 237.931 \text{ THz}.$$

Thus, the bandwidth of the T-band in frequency is given by

$$299.792 \text{ THz} - 237.931 \text{ THz} = 61.861 \text{ THz}.$$

By dividing the bandwidth by the bandwidth of each channel (100 GHz), we get

$$61.861 \text{ THz}/100 \text{ GHz} = 618.61,$$

so that the total number of channels is equal to 618.

2.6 S-, C-, and L-bands form a wide optical band whose edges are 1460 nm and 1625 nm in wavelength. As is the case in the previous solution, the edges in frequency are 205.3 and 184.5 THz, so that the total bandwidth in frequency is 20.8 THz. The number of optical channels is given by

$$\frac{20.8 \times 10^{12}}{50 \times 10^9} = 416,$$

where the bandwidth of each channel is 50 GHz. The bit rate of each channel is

$$C = 40 \times 10^9 \times \log_2 16 = 200 \times 10^9 = 160 \text{ Gb/s},$$

where the baud rate is 40 Gbaud and the number of symbols is 16. The total transmission capacity is

$$160 \text{ Gb/s} \times 416 = 66.56 \text{ Tb/s}.$$

Problems of Chap. 3

3.1 By substituting $J = 2J_{th}$ to (3.1), we get

$$f_r = K_r.$$

For the relaxation oscillation frequency of $5f_r$, (3.1) can be rewritten as

$$5f_r = K_r\sqrt{\frac{J}{J_{th}} - 1}.$$

By substituting $f_r = K_r$ to this equation, we get

$$\sqrt{\frac{J}{J_{th}} - 1} = 5,$$

so that

$$\frac{J}{J_{th}} - 1 = 25.$$

Finally, we get $J = 26J_{th}$.

3.2 By using (3.1), the relaxation oscillation frequency of f_r can be calculated as a function of J_{th}, for a fixed J, as follows:

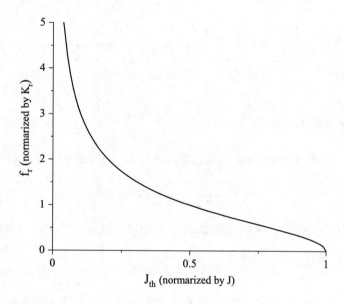

3.3 As of the previous solution, f_r can be calculated as a function of J, for a fixed J_{th}, as follows:

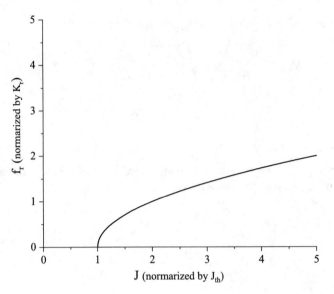

Problems of Chap. 4

4.1 The electric field amplitude along the c-axis E_3 is given by

$$E_3 = \frac{10.0}{100 \times 10^{-6}} = 10.0 \times 10^4.$$

By using (4.4), we can obtain the refractive index change along the c-axis as follows:

$$\Delta n_3 = -\frac{n_{0e}^3}{2} r_{33} E_3,$$

where n_3 and E_3 denote the refractive index and the electric field amplitude along the c-axis (z-axis). By substituting $E_3 = 10.0 \times 10^4$, $n_{0e} = 2.143$ and $r_{33} = 30.8 \times 10^{-12}$ into this equation, we get the refractive index change Δn_3, as follows:

$$\Delta n_3 = -\frac{2.143^3}{2} \times 30.8 \times 10^{-12} \times 10.0 \times 10^4 = 3.03 \times 10^{-5}.$$

4.2 The propagation delay can be calculated by (4.9). For the extraordinary ray, $n = 2.143$, so that

$$t_{DL} = \frac{2.143 \times 5.00 \times 10^{-2} \text{ m}}{299792458 \text{ m/s}} = 3.57 \times 10^{-10} \text{ s} = 357.4 \text{ ps}.$$

For the extraordinary ray, where $n = 2.223$,

$$t_{DL} = \frac{2.223 \times 5.00 \times 10^{-2} \text{ m}}{299792458 \text{ m/s}} = 370.8 \text{ ps}.$$

Thus, the delay difference is

$$370.8 - 357.4 \simeq 13 \text{ ps}.$$

4.3 As shown in Fig. 4.8, the zero-to-peak voltage is $V_{\pi PM}$, so that the root-mean-square value (effective value) of the signal voltage is

$$\frac{V_{\pi PM}}{\sqrt{2}}.$$

For a transmission line with a characteristic impedance denoted by Z_0, the signal power is

$$P_\pi = \left(\frac{V_{\pi PM}}{\sqrt{2}}\right)^2 \frac{1}{Z_0}.$$

When the characteristic impedance $Z_0 = 50 \ \Omega$,

$$P_\pi = \frac{V_{\pi PM}^2}{100}.$$

When $V_{\pi PM} = 5 \ V$,

$$P_\pi = \frac{5^2}{100} \text{ W} = 250 \text{ mW}.$$

4.4 The cutoff frequency described by (4.31) is

$$f_T = \frac{c}{2|n_m - n_0|L}$$

$$= \frac{299792458 \text{ m/s}}{2 \times (\sqrt{28.0} - 2.143) \times 2.00 \times 10^{-2} \text{ m}}$$

$$= 2.38 \times 10^9 \text{ Hz} = 2.38 \text{ GHz}.$$

If we use (4.32), the cutoff frequency is

$$f_T = 0.6 \frac{c}{|n_m - n_0|L} = 2.86 \text{ GHz}.$$

4.5 By substituting (4.47) to (4.46), we get

$$\Phi(t) = \alpha_0 g(t),$$

so that

$$\frac{d\Phi}{dt} = \alpha_0 \frac{dg(t)}{dt}.$$

The derivative of E is

$$\frac{dE}{dt} = -E_0 \frac{dg(t)}{dt} \sin[g(t)].$$

Thus,

$$\frac{d\Phi}{dt} \bigg/ \frac{1}{E}\frac{dE}{dt} = E\frac{d\Phi}{dt}\left(\frac{dE}{dt}\right)^{-1}$$

$$= E_0 \cos[g(t)]\,\alpha_0 \frac{dg(t)}{dt}\left(-E_0\frac{dg(t)}{dt}\sin[g(t)]\right)^{-1}$$

$$= -\alpha_0 \cot[g(t)].$$

4.6 As shown in (4.57), the imbalance η can be expressed by

$$\eta = \frac{2}{\sqrt{ER}}.$$

When the ER is equal to 60 dB,

$$\eta = \frac{2}{\sqrt{10^6}} = 2 \times 10^{-3}.$$

Problems of Chap. 5

5.1 Required voltage for 360° phase shift is double the half-wave voltage. The symbol separation in angle is $360/N$, where N is the number of symbols, so that the voltage step for 16-level PSK is

$$8.0\,\text{V} \times 2/16 = 1.0\,\text{V}.$$

5.2 From (5.6), we get

$$N_{\text{MZ}} = \log_2 N_{\text{QAM}}.$$

Thus, the number of the required MZIs is

$$\log_2 1024 = 10.$$

Problems of Chap. 6

6.1 For $J_0(z)$, by using (6.8), we get

$$J_0(z) = \sum_{m=0}^{2} \frac{(-1)^m}{m!m!} \left(\frac{z}{2}\right)^{2m}$$

$$= 1 - \frac{z^2}{2^2} + \frac{z^4}{2 \times 2 \times 2^4}$$

$$= 1 - \frac{z^2}{4} + \frac{z^4}{64}.$$

Similarly, for $J_1(z)$,

$$J_1(z) = \sum_{m=0}^{1} \frac{(-1)^m}{m!(m+1)!} \left(\frac{z}{2}\right)^{2m+1}$$

$$= \frac{z}{2} - \frac{z^3}{2 \times 2^3}$$

$$= \frac{z}{2} - \frac{z^3}{16}.$$

For $J_2(z)$,

$$J_2(z) = \sum_{m=0}^{1} \frac{(-1)^m}{m!(m+2)!} \left(\frac{z}{2}\right)^{2m+2}$$

$$= \frac{z^2}{2 \times 4} - \frac{z^4}{6 \times 2^4}$$

$$= \frac{z^2}{8} - \frac{z^4}{96}.$$

6.2 From (6.62), we get

$$\left| \sum_{n=-\infty}^{\infty} \left[J_n(A_i) e^{in\omega_m t + in\phi_i} \right] \right| = \left| e^{iA_i \sin(\omega_m t + \phi_i)} \right| = 1.$$

Thus,

$$\left| \sum_{n=-\infty}^{\infty} \left[J_n(A_i) e^{in(\omega_m t + \phi_i)} \right] \right|^2$$

$$= \sum_{n=-\infty}^{\infty} \left[J_n(A_i) e^{in(\omega_m t + \phi_i)} \right] \sum_{n=-\infty}^{\infty} \left[J_n(A_i) e^{-in(\omega_m t + \phi_i)} \right] = 1.$$

This equation can be rewritten as,

$$\sum_{n=-\infty}^{\infty} [J_n(A_i)]^2 + \sum_{n,n'=-\infty(n\neq n')}^{\infty} J_n(A_i)J_{n'}(A_i)e^{i(n-n')(\omega_m t+\phi_i)}$$

$$= \sum_{n=-\infty}^{\infty} [J_n(A_i)]^2 + \sum_{n,n'=-\infty(n\neq n')}^{\infty} J_n(A_i)J_{n'}(A_i)\cos\left[(n-n')(\omega_m t+\phi_i)\right] = 1.$$

Here, we consider an integral of the left side of this equation over a period of the modulating signal, as follows:

$$\int_0^T \sum_{n=-\infty}^{\infty} [J_n(A_i)]^2 \, dt$$

$$+ \int_0^T \sum_{n,n'=-\infty(n\neq n')}^{\infty} J_n(A_i)J_{n'}(A_i)\cos\left[(n-n')(\omega_m t+\phi_i)\right] dt$$

$$= \sum_{n=-\infty}^{\infty} [J_n(A_i)]^2 \int_0^T dt$$

$$+ \sum_{n,n'=-\infty(n\neq n')}^{\infty} J_n(A_i)J_{n'}(A_i) \int_0^T \cos\left[(n-n')(\omega_m t+\phi_i)\right] dt$$

$$= \sum_{n=-\infty}^{\infty} [J_n(A_i)]^2 \, T,$$

where the period is given by $T = 2\pi/\omega_m$.

The integral of the right side of the equation is equal to T. Thus, we get

$$\sum_{n=-\infty}^{\infty} [J_n(A_i)]^2 = 1.$$

6.3

$$P_{t2}^{(2)} = \left(1 - \frac{A^2}{4}\right)^2 + 2 \times \left(\frac{A}{2}\right)^2 + 2 \times \left(\frac{A^2}{8}\right)^2$$

$$= 1 - \frac{A^2}{2} + \frac{A^4}{16} + \frac{A^2}{2} + \frac{A^4}{32} = 1 + \frac{3A^4}{32}$$

6.4 By using the solutions of problem 6.1 and (6.15) and (6.16), the sideband components up to the fourth order can be expressed by expansions using the fourth order of A. By substituting them into

$$P_{t4}^{(4)} = \{J_0(A)\}^2 + 2\sum_{n=1}^{4} \{J_n(A)\}^2,$$

we get

$$P_{t4}^{(4)} = \left(1 - \frac{A^2}{4} + \frac{A^4}{64}\right)^2 + 2 \times \left(\frac{A}{2} - \frac{A^3}{16}\right)^2 + 2 \times \left(\frac{A^2}{8} - \frac{A^4}{96}\right)^2$$

$$+ 2 \times \left(\frac{A^3}{48}\right)^2 + 2 \times \left(\frac{A^4}{384}\right)^2$$

$$= 1 - \frac{5A^6}{1152} + \frac{35A^8}{73728}.$$

6.5

$$P_n = \frac{J_n^2(A)}{4} \left|(-1)^n + e^{i(\phi_B + n\phi)}\right|^2$$

$$= \frac{J_n^2(A)}{4} \left|e^{-in\pi} + e^{i(\phi_B + n\phi)}\right|^2$$

$$= \frac{J_n^2(A)}{4} \left|e^{i(\phi_B + n\phi)}(e^{-in\pi}e^{-i(\phi_B + n\phi)} + 1)\right|^2$$

$$= \frac{J_n^2(A)}{4} \left|e^{i(\phi_B + n\phi)}\right|^2 \left|e^{-in\pi}e^{-i(\phi_B + n\phi)} + 1\right|^2$$

$$= \frac{J_n^2(A)}{4} \left|e^{-i[\phi_B + n(\phi + \pi)]} + 1\right|^2$$

$$= \left\{J_n(A) \cos \frac{\phi_B + n(\phi + \pi)}{2}\right\}^2$$

$$= \frac{J_n^2(A)}{2} \left[\cos\{\phi_B + n(\phi + \pi)\} + 1\right]$$

$$= \frac{J_n^2(A)}{2} \left[(-1)^n \cos(\phi_B + n\phi) + 1\right]$$

$$= \frac{J_n^2(A)}{2} \left[(-1)^n \cos \Phi + 1\right].$$

Problems of Chap. 7

7.1 By using (6.12), (6.13) and (6.14), we get

$$E_{2N}/E_{1N} = \alpha_0 \left[A - 2 \times \frac{A^2}{8} \left(\frac{A}{2}\right)^{-1} + A \times \frac{A^2}{8}\right] \simeq \alpha_0 \left(A - \frac{A}{2}\right) = \frac{\alpha_0 A}{2}.$$

7.2 By using (6.33), we get

$$J_3'(A) = J_2(A) - \frac{3J_3(A)}{A}$$

$$J_1'(A) = J_0(A) - \frac{J_1(A)}{A}.$$

By substituting them into (7.42), we can derive (7.43), as follows:

$$\frac{E_{3F}}{E_{2F}} = \frac{\alpha_0 A \left[J_2(A) - \frac{3J_3(A)}{A} - \left(J_0(A) - \frac{J_1(A)}{A} \right) \frac{J_3(A)}{J_1(A)} \right]}{J_2(A)}$$

$$= \alpha_0 \frac{AJ_1(A)J_2(A) - 2J_1(A)J_3(A) - AJ_0(A)J_3(A)}{J_1(A)J_2(A)},$$

where $E_{2F} = J_2(A)$ as described in (7.36).

7.3 By using (6.12), (6.13), (6.14), and (6.15), E_{3F}/E_{2F} can be expressed by

$$E_{3F}/E_{2F} = \alpha_0 \frac{A \times \frac{A}{2} \times \frac{A^2}{8} - 2 \times \frac{A}{2} \times \frac{A^3}{48} - A \times \frac{A^3}{48}}{\frac{A}{2} \times \frac{A^2}{8}}$$

$$= \alpha_0 A \left(\frac{1}{16} - \frac{1}{48} - \frac{1}{48} \right) \bigg/ \frac{1}{16} = \frac{\alpha_0 A}{3}.$$

7.4 By using (7.61) and (7.62), we can derive (7.64), as follows:

$$R = \frac{e^{i\omega_0 t}}{2} \sum_{k=0}^{\infty} J_{2k+1}(A_m) \left[\left\{ i \cdot (-1)^k e^{if_{FSK}(t)} + e^{-if_{FSK}(t)} \right\} e^{i(2k+1)\omega_m t} \right.$$

$$\left. + \left\{ i \cdot (-1)^k e^{if_{FSK}(t)} - e^{-if_{FSK}(t)} \right\} e^{-i(2k+1)\omega_m t} \right]$$

$$= \frac{e^{i[\omega_0 t + \pi/4]}}{2} \sum_{k=0}^{\infty} J_{2k+1}(A_m) \left[\left\{ (-1)^k e^{i[f_{FSK}(t)+\pi/4]} + e^{-i[f_{FSK}(t)+\pi/4]} \right\} e^{i(2k+1)\omega_m t} \right.$$

$$\left. + \left\{ (-1)^k e^{i[f_{FSK}(t)+\pi/4]} - e^{-i[f_{FSK}(t)+\pi/4]} \right\} e^{-i(2k+1)\omega_m t} \right]$$

$$= e^{i[\omega_0 t + \pi/4]} \sum_{k=0}^{\infty} \left[J_{4k+1}(A_m) \left\{ \cos\left[f_{FSK}(t) + \pi/4 \right] e^{i(4k+1)\omega_m t} \right. \right.$$

$$\left. + i \sin\left[f_{FSK}(t) + \pi/4 \right] e^{-i(4k+1)\omega_m t} \right\}$$

$$- J_{4k+3}(A_m) \left\{ i \sin\left[f_{FSK}(t) + \pi/4 \right] e^{i(4k+3)\omega_m t} \right.$$

$$\left. \left. + \cos\left[f_{FSK}(t) + \pi/4 \right] e^{-i(4k+3)\omega_m t} \right\} \right]$$

$$
= e^{i[\omega_0 t + \pi/4]} \Bigg[\cos[f_{FSK}(t) + \pi/4] \sum_{k=0}^{\infty} (-1)^k J_{2k+1}(A_m) e^{i(-1)^k (2k+1)\omega_m t}
$$

$$
+ i \sin[f_{FSK}(t) + \pi/4] \sum_{k=0}^{\infty} (-1)^k J_{2k+1}(A_m) e^{-i(-1)^k (2k+1)\omega_m t} \Bigg]
$$

$$
= e^{i[\omega_0 t + \pi/4]} \Bigg[\cos[f_{FSK}(t) + \pi/4] \left\{ J_1(A_m) e^{i\omega_m t} - J_3(A_m) e^{-i3\omega_m t} \right.
$$

$$
\left. + J_5(A_m) e^{i5\omega_m t} - J_7(A_m) e^{-i7\omega_m t + \cdots} \right\}
$$

$$
+ i \sin[f_{FSK}(t) + \pi/4] \left\{ J_1(A_m) e^{-i\omega_m t} - J_3(A_m) e^{i3\omega_m t} \right.
$$

$$
\left. + J_5(A_m) e^{-i5\omega_m t} - J_7(A_m) e^{i7\omega_m t + \cdots} \right\} \Bigg].
$$

Problems of Chap. 8

8.1

The power of the modulating signal is 400 mW (= 26 dBm). When the characteristic impedance is 50 Ω, the zero-to-peak amplitude can be derived from

$$
\left(\frac{V_{0p}}{\sqrt{2}} \right)^2 \frac{1}{50\ \Omega} = 0.40\ \text{W},
$$

so that

$$
V_{0p} = \sqrt{40} = 6.33.
$$

From (8.1), we get

$$
V_{\pi MZM(RF)} = \frac{3.14 \times 6.33\ \text{V}}{1.60\ \text{rad} + 1.60\ \text{rad}} = 6.2\ \text{V}.
$$

8.2 The half-wave voltage of each phase modulator is double the half-wave voltage of the zero-chirp MZM. Thus, we get,

$$
V_{\pi PM(RF)} = 2 V_{\pi MZM(RF)} = 12.4\ \text{V}.
$$

8.3 The power of the modulating signal is 200 mW (= 23 dBm). As with problem 8.1, the zero-to-peak amplitude is given by

$$
V_{0p} = \sqrt{20} = 4.47.
$$

When $\alpha_0 = 1$, the optical phase shift at the second phase modulator whose induced phase is denoted by $v_2(t)$ in (4.34) is equal to zero, as shown in (4.49). Thus, the total optical phase difference between the two arms is equal to 1.60 rad, where $A_1 = 2.4$ and $A_2 = 0$. From (8.1), we get

$$V_{\pi\text{MZM(RF)}} = \frac{3.14 \times 4.47 \text{ V}}{1.60 \text{ rad}} = 8.8 \text{ V}.$$

Index

© Springer Nature Switzerland AG 2022
T. Kawanishi, *Electro-optic Modulation for Photonic Networks*, Textbooks in
Telecommunication Engineering, https://doi.org/10.1007/978-3-030-86720-1

Printed in the United States
by Baker & Taylor Publisher Services